SWITCHING IN
SEMICONDUCTOR
DIODES

MONOGRAPHS IN SEMICONDUCTOR PHYSICS

Volume 1:
Heavily Doped Semiconductors
by Viktor I. Fistul'

Volume 2:
Liquid Semiconductors
by V. M. Glazov, S. N. Chizhevskaya, and N. N. Glagoleva

Volume 3:
Semiconducting II-VI, IV-VI, and V-VI Compounds
by N. Kh. Abrikosov, V. F. Bankina, L. V. Poretskaya,
L. E. Shelimova, and E. V. Skudnova

Volume 4:
Switching in Semiconductor Diodes
by Yu. R. Nosov

In preparation:
Semiconducting Lead Chalcogenides
by Yu. I. Ravich, B. A. Efimova, and I. A. Smirnov

Organic Semiconductors and Biopolymers
by L. I. Boguslavskii and A. V. Vannikov

SWITCHING IN SEMICONDUCTOR DIODES

Yurii R. Nosov

A. F. Ioffe Physicotechnical Institute
Leningrad, USSR

Translated from Russian by
Albin Tybulewicz
Editor, *Soviet Physics-Semiconductors*

 Springer Science+Business Media, LLC • 1969

Yurii Romanovich Nosov was born in 1931, in Moscow. He graduated in 1954 from the Physics Department of the M. V. Lomonosov State University in Moscow. Since graduation, he has been engaged in research on, and the design of, fast-response semiconductor pulse diodes. In 1964, Nosov was awarded the degree of Candidate of Technical Sciences by the A. F. Ioffe Physicotechnical Institute in Leningrad. In 1965, he was promoted to the rank of senior scientist in the specialty of electronic technology and devices. He is a member of the All-Union Society "Znanie," the A. S. Popov Scientific and Technical Society for Radio Engineering and Electrical Communications, and of the "Trud" Sporting Association.

The original Russian text was published by Nauka Press, Moscow, in 1968 as part of a series on "Physics of Semiconductors and Semiconducting Devices" and has been revised by the author for this English edition.

Юрий Романович Носов

ФИЗИЧЕСКИЕ ОСНОВЫ РАБОТЫ ПОЛУПРОВОДНИКОВОГО ДИОДА
В ИМПУЛЬСНОМ РЕЖИМЕ

FIZICHESKIE OSNOVY RABOTY POLUPROVODNIKOVOGO DIODA
V IMPUL'SNOM REZHIME

ISBN 978-1-4899-6172-3   ISBN 978-1-4899-6343-7 (eBook)
DOI 10.1007/978-1-4899-6343-7

Library of Congress Catalog Card Number 69-12535

Preface to the American Edition

It gives me great pleasure to learn that this book, whose origin owes much to the work of American scientists and engineers on semiconductor technology, will reach American and other English-speaking readers.

I am grateful to Plenum Publishing Corporation for arranging the American edition of this book and to Mr. Albin Tybulewicz for his translation.

September 5, 1968   Yu. R. Nosov

Preface to the American Edition

Preface to the Russian Edition

One of the most important applications of semiconductor diodes is their use in electronic pulse circuits.

The response of these diodes under switching conditions is governed by the phenomena of accumulation and dispersal of non-equlibrium carriers, which are also observed in other p-n junction devices.

It was found in the late 1940's that when point-contact germanium diodes were used in circuits through which short (several tenths of a microsecond) electrical pulses were being passed, the effective reverse resistance of these diodes decreased considerably below the static value. Further studies showed that when a diode was switched rapidly from the forward to the reverse direction, an anomalously large reverse current flowed for some time.

In view of the importance of this phenomenon in the efforts to reduce the response time of pulse circuits, many investigations of the phenomenon were carried out and these investigations provided the basis of a theory of transient processes in semiconductor diodes.

It was established that the rate of dispersal of nonequilibrium carriers, accumulated in the diode base, is governed by the properties of the diode itself, as well as by the switching conditions. This made it possible to reduce the response time of the diodes and to develop methods for the determination of the optimal operating conditions of these devices in pulse circuits. Later it was found that this theory of transient processes had other important applications, in addition to its usefulness in reducing the response time of diodes and of pulse diode circuits.

This theory has served as the corner-stone of the studies of
the processes of the accumulation and dispersal of excess charges
in transistors and thyristors and, through its application, the upper
frequency limit of these devices has been increased. Secondly, the
theory has stimulated the development of such new devices as
charge-storage diodes and some special types of varactor for frequency multiplication.

The development of the theory of transient processes in diodes
has made it possible condiderably to extend investigations of relaxation processes in semiconductors under nonequilibrium conditions, in particular, the recombination of nonequilibrium carriers.

The main purpose of this book is to acquaint the reader with
the physical basis of the operation of a semiconductor diode under
pulse conditions, and to give, whenever possible, quantitative relationships describing transients in diodes.

The variety of various possible switching conditions in circuits is practically infinite and, therefore, it has not been possible
to discuss all the conditions. Instead, a description is given of the
most general mathematical methods for solving the equations describing transient processes, and these methods are presented in
such a way as to enable the reader to analyze those pulse conditions
which are not discussed in this book.

A model of a planar diode with a semi-infinite base is discussed in the greatest detail. This is done because the basic formulas for this model are the simplest and clearest; moreover, the
most important relationships obtained for the planar diode are
found to apply approximately to other diodes.

Problems associated with the application of the theory of
transient processes in the practical design of fast-response diodes
and of circuits incorporating such diodes are not discussed in this
book; readers interested in the design aspects are directed to spe-
cialist texts on this subject [62, 103, 166].

The list of references is reasonably complete but it does not
cover fully all the numerous applications.

For convenience, the symbols used are explained in the text
and are also collected together in a separate list.

The author is grateful to Candidate of Technical Sciences L. S. Berman for reading the manuscript and for his valuable advice; to Candidate of Technical Sciences O. K. Mokeev for discussing some of the problems condidered in this book; and to L. V. Gubyrin and L. A. Kuranova for their help in writing Chapter III and selecting the literature relevant to this chapter. The author is much indebted to I. S. Egorova without whose help in the preparation of the manuscript this book would not have been published.

Contents

Notation xiii

CHAPTER I. Basic Electronics of the Switching
Processes in Semiconductor p–n Junctions 1

§1. Introduction 1
§2. Transformation of Basic Equations 4
§3. Solution of the Diffusion Equation (at Low
     Injection Levels) 10

CHAPTER II. Switching in a Planar Diode 25

§4. Transient Processes without a Limiting
     Resistance in the Diode Circuit 25
§5. Switching of a Diode Circuit with a
     Limiting Resistance 38
§6. Switching of a Diode Circuit with an
     Infinite Resistance 53
§7. Small–Signal Transient Characteristics
     of a Diode 66
§8. Methods for the Observation of Transient
     Processes in Diodes 69
§9. Main Experimental Results 82

CHAPTER III. Planar Diode with a Thin Base 96

§10. Steady–State Distribution of Holes in
      the Base 96
§11. Switching without a Resistance in the
      Diode Circuit 107
§12. Switching in a Circuit with a Limiting
      Resistance 118
§13. General Estimate of the Response
      of a Thin–Base Diode 127

CHAPTER IV. Transient Processes in a Diode
with a Small–Area Rectifying Contact   130

§14. Ideal Model of a Point-Contact Diode   131
§15. Transient Conditions   141
§16. Experimental Investigations   153

CHAPTER V. Effect of an Electric Field in a
Diode Base on Transient Processes   160

§17. Built-in Internal Field in a Diode Base   160
§18. Forward-Biased Diode with a Built-in
     Field   165
§19. First (Recovery) Phase   171
§20. Reverse Current Decay   175

CHAPTER VI. Transient Processes in Diodes
During the Passage of a Forward Current
Pulse   179

§21. Introduction   179
§22. Establishment of a Forward Resistance
     in a Planar Diode   183
§23. Establishment of a Forward Voltage
     Across a Diode with a Hemispherical
     p–n Junction.   191

CHAPTER VII. Transient Processes in
Semiconductor Diodes and Fundamentals
of Recombination Theory   197

§24. Introduction.   197
§25. Lifetime of Holes under Various
     Recombination Conditions   201
§26. Influence of Trapping Levels on
     Transient Processes in Diodes   209
§27. Recombination Properties of Gold-Doped
     Germanium and Silicon   214

Literature Cited   226

Index   231

Notation

$A = r_0/L_p$	normalized radius of a hemispherical p-n junction
$B = i_0/i_f$	ratio of the current during the first phase of a reverse switching transient and the forward current
$b = \mu_n/\mu_p$	ratio of the electron and hole mobilities
C_e, C_{p-n}, C_d	capacitance of, respectively, the encapsulation, the p-n junction, and the diode as a whole
D_n, D_p	diffusion coefficients of electrons and holes
E	electric field
$E_n = qL_pE/2kT$	normalized electric field
E_t	energy level of a trap
E_v, E_c	energy levels of the upper edge of the valence band and the lower edge of the conduction band
E_α, E_β	energy levels of α and β trapping centers
h_K	transfer function of the bias ratio
h_{p-n}	width of the space-charge region
h_Y	transfer function of the circuit admittance
i_{rec}	reverse current corresponding to the completion of the second phase of a reverse switching transient (recovery current)
i, j	total diode current and diode current density
i_f, j_f	forward diode current and forward current density
i_0, j_0	reverse diode current and reverse current density during the first phase of a switching transient

j_n, j_p	electron and hole current densities in a diode
i_s, j_s	saturation current and saturation current density of a p-n junction
k	Boltzmann's constant
L_n	diffusion length of electrons in a p-type region
L_p	diffusion length of holes in an n-type region
l_D	Debye screening length
$l_p = L_p\sqrt{2b/(b+1)}$	effective diffusion length of holes at high injection levels
N_c, N_v	effective densities of states in conduction and valence bands
N_d, N_a	donor and acceptor concentrations
N_t	trap concentration
N_α, N_β	densities of α and β trapping levels
n, p	electron and hole densities
n_{n0}, n_{p0}	densities of electrons in n- and p-type semiconductors under thermodynamic equilibrium conditions
P_{n0}, P_{p0}	densities of holes in n- and p-type semiconductors under thermodynamic equilibrium conditions
p_1	density of holes in the base at the edge of the base side of the space-charge region during passage of a steady-state forward current ("impressed density")
Q_{rec}	recovered charge
Q_{st}	stored charge
q	magnitude of electron charge
$R = r/L_p$	normalized radius vector
R_b	base resistance
R_i	differential forward resistance of a diode
R_l	load resistance in a diode circuit
R_R	"residual" base resistance
r_0	radius of a hemispherical p-n junction
S	p-n junction area
S_R	surface recombination velocity
T	absolute temperature
$\mathcal{J} = t/\tau_p$	normalized time
$\mathcal{J}_d = t_d/\tau_p$	normalized delay time between the end of a forward current pulse and beginning of a reverse voltage pulse

$\mathcal{J}_f = t_f/\tau_p$	normalized duration of a forward current pulse
$\mathcal{J}_l = t_l/\tau_p$	normalized duration of the linear part of the "tail" of the postinjection (open-circuit) emf across a p-n junction
$\mathcal{J}_{pf} = t_{pf}/\tau_p$	normalized pulse front rise time
$\mathcal{J}_1 = t_1/\tau_p$	normalized duration of the first phase of a transient after switching from forward to reverse direction
$\mathcal{J}_2 = t_2/\tau_p$	normalized duration of the second phase of a transient after switching from forward to reverse direction (decay of the reverse current from i_0 to $0.1\, i_0$)
t_C	time constant of charging the barrier capacitance
U_r	amplitude of a reverse voltage pulse (step)
u_b	voltage drop across the base
u_d	total voltage drop across a diode
u_D	Dember voltage drop across the base
u_f	forward voltage drop across a diode
u_0	ohmic voltage drop across the base
u_{p-n}	voltage drop across a p-n junction
u_r	steady reverse voltage
v_d	drift velocity of carriers
v_n, v_p	thermal velocities of electrons and holes
W	base thickness
$W_n = W/L_p$	normalized base thickness
$X = x/L_p$	normalized position coordinate
γ	injection efficiency of a p-n junction
$\Delta = P_1/n_{n0}$	injection level
δ_n, δ_p	excess electron and hole densities
θ_M	dielectric (Maxwellian) relaxation time
μ_n, μ_p	electron and hole mobilities
ρ_n	resistivity of an n-type semiconductor
σ_n, σ_p	cross sections for the capture of an electron and a hole by a recombination center
τ_n, τ_p	electron and hole lifetimes
$\tau_0, \tau(\Delta), \tau_\infty$	lifetimes of minority carriers, respectively, at low, arbitrary, and high injection levels
τ_{n0}	lifetime of electrons in a heavily-doped p-type semiconductor

τ_{p0} lifetime of holes in a heavily-doped n-type
                                   semiconductor

τ_p^{sts}, τ_p^{nst} steady-state and non-steady-state hole life-
                                   times

τ_{rad} radiative recombination lifetime

τ_t average carrier trapping time

φ_0 equilibrium barrier height (built-in potential)
                                   in a p-n junction

Chapter I

Basic Electronics of the Switching Processes in Semiconductor p-n Junctions

§ 1. INTRODUCTION

The operation of a semiconductor diode under pulse conditions is characterized by rapid and frequent transitions from one steady state to another. A state of a dynamic system (such as a diode with other circuit elements connected to it) is called steady (stationary) if all the variables describing its behavior do not vary with time or are periodic functions of time (within a defined time interval) [1].

Pulses applied to a diode are sudden jumps of the electric voltage or current. Two types of pulse can be distinguished: video pulses and radio pulses. In a video pulse, a rapid rise of the electric signal from zero to its maximum is followed (after a certain time interval) by its decay back to zero. A radio pulse is a brief packet of high-frequency oscillations (these oscillations are known as the carrier signal); the duration of a radio pulse is at least several times longer than the carrier-signal period.

In the majority of cases, the studies of fast-response semiconductor diodes are concerned with video pulses; therefore, for brevity, we shall simply call them pulses. Unless otherwise stated, the pulse shape will be assumed to be perfectly rectangular. When a "pulse" is defined in this way, a steady state of the diode can be represented by constant values of the variables describing its behavior.

It is known that the rapid transition of a system or a circuit element from one steady state to another is accompanied by transient processes. The state of a diode which cannot be described

1

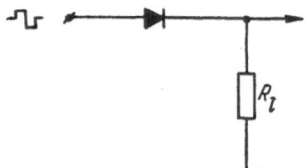

Fig. 1.1. Switching of a diode
from the forward to the reverse
direction.

by time–independent values of its parameters is called a transient
(nonstationary) state.

To understand the physics of the behavior of a diode transient
conditions, we must consider systems with the smallest number of
elements, apart from the diode. The simplest system is a series
circuit consisting of a diode and a load resistance (Fig. 1.1). Then,
an investigation of the behavior of a diode under pulse conditions
reduces to an analysis of the transient processes in such a system.

The most complete analysis of transient processes can be
carried out for linear systems. In such cases, we can use the
principle of superposition, which makes it possible to use a transfer function or characteristic. The transfer function is some function $h(t)$ which describes the variation of the output signal when the
input signal increases suddenly from zero to some constant value
which is assumed to be unity. In other words, the transfer function
is the reaction of a system to a step function.

If the transfer function of a linear system is known, we can
use the Duhamel integral to determine the response $y(t)$ to an arbitrary input signal:

$$y(t) = \int_0^t h(t-\theta)\, df(\theta) \qquad (1.1)$$

where θ is the integration variable, $df(\theta)$ is the differential representing the effect of an arbitrary input function $f(t)$ in which the
argument t is replaced with θ.

It is known that the current-voltage characteristic of a semiconductor diode is strongly nonlinear, particularly when the diode
is biased in the forward direction. Only at very low values of the
change in the external voltage, when the condition

$$\Delta u \ll \frac{kT}{q} \qquad (1.2)$$

is obeyed (here, k is the Boltzmann constant, T is the absolute temperature, q is a positive charge equal in its absolute magnitude to the electron charge), can we use a linear approximation in which it is permissible to replace a diode with a resistance equal to the differential resistance R_i at the relevant voltage.* Calculations show that this differential resistance depends on the frequency of the external signal; moreover, the equivalent circuit of the diode must include a capacitor whose capacitance also depends on the frequency. This is why transients are observed when the voltage across a diode is changed suddenly.

The fundamentals of the frequency characteristics of semiconductor diodes have been given in the now classical paper of Shockley [2]; they have been developed by Kalashnikov and Penin [3] and by Gossick [4]; the most complete and systematic analysis of small-signal transient characteristics has been carried out by Adirovich [5].

The condition given by Eq. (1.2) is obeyed in practice only under very special conditions. A much more typical situation in the pulse operation of semiconductor devices is described by an inequality which is the reverse of that given by Eq. (1.2). Diodes are used in pulse circuits as basically nonlinear elements, in which the current is not proportional to the applied voltage. Usually, the diode resistance is very high for one polarity of the external voltage while for the opposite polarity it is negligibly small. This case is described as the switching regime.

We shall investigate the transient processes which accompany the switching of a diode.

In accordance with the various physical phenomena which govern the nature of a given nonstationary state, we can divide the various transient processes in diodes into two groups:

1. The processes associated with the switching of a diode from the neutral (or reverse) to the forward direction. These processes consist basically of the injection of minority carriers, the speading of these carriers into regions far from the p-n junction, and the accumulation of these carriers in such regions.

---

*This approach is known as the small-signal approximation.

2. The processes associated with the switching of a diode from the forward to the reverse (or neutral) direction. In this case, the rate of transition from one steady state to another is governed by the recombination of accumulated excess carriers and their extraction by a p-n junction.*

The switching processes in semiconductor diodes are investigated in the following sequence: an ideal diode model is selected; a differential equation describing the behavior of the carriers in the active parts of the diode is derived; the initial distribution of the electrons and holes is determined, and the boundary conditions at the p-n junction and at the ohmic contacts are satisfied; the differential equation is solved for mobile carriers; the time dependences of the current through the diode and the voltage across it are determined.

The theoretical equations obtained in this way are then checked by suitable experiments.

§ 2. TRANSFORMATION OF BASIC EQUATIONS

We shall base our analysis on a model of an infinite semiconductor with a planar p-n junction, i.e., we shall solve a one-dimensional problem. To make this problem more specific, we shall assume that the electrical conductivity of the semiconductor is considerably higher in the p-type region than in the n-type region and that both regions are uniform in their electrical properties. We shall call the heavily doped part of the semiconductor the emitter and lightly doped part the base. The n- and p-type parts of a crystal, together with the ohmic contacts, represent our semiconductor diode.

The behavior of the carriers in the base under steady-state and transient conditions is described by the following system of equations:†

$$j_p = q\mu_p p E - q D_p \frac{\partial p}{\partial x},$$ (1.3)

---

* The phenomena of injection and extraction of minority carriers by a p-n junction, which govern the operation of a transistor, have been considered in detail in many papers and monographs (cf., for example, [6, 7]); we shall not discuss these phenomena in detail.
† The equations considered in this book are based on the instantaneous Fick's law, which postulates that the diffusion flux of carriers is proportional to the gradient of their

$$j_n = q\mu_n nE + qD_n \frac{\partial n}{\partial x}, \tag{1.4}$$

$$j = j_p + j_n, \tag{1.5}$$

$$\frac{\partial p}{\partial t} = -\frac{p - p_{n0}}{\tau_p} - \frac{1}{q}\frac{\partial j_p}{\partial x}, \tag{1.6}$$

$$\frac{\partial n}{\partial t} = -\frac{n - n_{n0}}{\tau_n} + \frac{1}{q}\frac{\partial j_n}{\partial x}, \tag{1.7}$$

$$\frac{\partial E}{\partial x} = \frac{4\pi q}{\varepsilon}(p - n - p_{n0} + n_{n0}). \tag{1.8}$$

In these equations, j_p, j_n, and j are the hole, electron, and total current densities, respectively; p, n, p_{n0}, and n_{n0} are the total and equilibrium densities of the holes and electrons in the n–type region; τ_p, τ_n, D_p, D_n, μ_p, and μ_n are the lifetimes, diffusion coefficients, and mobilities of the holes and electrons; ε is the permittivity; E is the electric field intensity at a given point in the base.

Equations (1.3) and (1.4) are simply the definitions of the hole and electron current densities as the sums of the drift and diffusion components; the expressions (1.6) and (1.7) are simply the equations of continuity for the holes and electrons; finally, Eq. (1.8) is Poisson's equation for a semiconductor with fully ionized donors and acceptors whose concentrations are, respectively, $N_d = n_{n0}$ and $N_a = p_{n0}$.

The transformation of the system of equations (1.3)–(1.8) yields nonlinear second–order differential equations for p(x, t) and n(x, t), which — in general — cannot be solved analytically.

Only in some special cases can the system (1.3)–(1.8) be made linear. The simplest best-known approximation is the low-injection-level case, when the following inequality is satisfied by the base

$$p(x, t) \ll n_{n0}. \tag{1.9}$$

---

density at any given moment. Estimates, obtained using the relationships given in [167], show that this assumption is justified in the case of germanium and silicon provided the transients last longer than 10^{-11} sec.

In this case, because the departure from the state of thermodynamic equilibrium is slight, we can assume that at each point in the base the condition of charge neutrality is obeyed, i.e.,

$$\delta p \equiv (p - p_{n0}) = \delta n \equiv (n - n_{n0}).$$ (1.10)

It follows from Eq. (1.8) that the electric field intensity is independent of the position coordinate, and Eqs. (1.3) and (1.6) yield

$$\frac{\partial p}{\partial t} = D_p \frac{\partial^2 p}{\partial x^2} - \mu_p E \frac{\partial p}{\partial x} - \frac{p - p_{n0}}{\tau_p}.$$ (1.11)

In the majority of cases of practical interest (when the conductivity of the n-type region is far from the intrinsic value), we may assume that the whole external voltage applied to a diode is concentrated in the p-n junction region and, consequently, the drift current of minority carriers can be neglected compared with the diffusion current. This approximation has been used by Shockley in the derivation of the basic relationships for a p-n junction [2] and by other workers who have developed the theory of transient processes in diodes [8-12]. In this case, Eq. (1.11) simplifies and becomes

$$\frac{\partial p}{\partial t} = D_p \frac{\partial^2 p}{\partial x^2} - \frac{p - p_{n0}}{\tau_p}.$$ (1.12)

According to the recombination theory of Shockley and Read [13] and of Hall [14], the lifetime of minority carriers at low injection levels is independent of their density and Eq. (1.12) can be transformed to

$$\frac{1}{D_p} \frac{\partial p}{\partial t} = \frac{\partial^2 p}{\partial x^2} - \frac{p - p_{n0}}{L_p^2},$$ (1.13)

where L_p is the diffusion length of holes in the base, which is independent of the coordinate and equal to $\sqrt{D_p \tau_p}$.

Thus, the assumption of a low injection level and a small drift current of minority carriers * allows us to described the be-

---

* We must stress particularly that we can neglect the drift current only in the case of minority carriers, because the majority-carrier drift current is considerable throughout the base, and far from the p-n junction, where $\partial n / \partial x \to 0$, the whole current through the diode is the drift current of the majority carriers.

havior of holes in the base by means of the diffusion equation with constant coefficients, which has a simple solution.

The system of equations (1.3)-(1.8) can be made linear also in the case of high injection levels, provided the following condition is satisfied:

$$\delta p \gg n_{n0}. \tag{1.14}$$

The transformation used by Rittner [15] is also based on the neutrality condition given by Eq. (1.10), from which it follows that, in particular, $\partial p / \partial x = \partial n / \partial x$ and $\partial p / \partial t = \partial n / \partial t$. Using these two equalities, we can reduce the system of equations (1.3)-(1.8) to

$$\frac{\partial p}{\partial t} = D_p \frac{\partial^2 p}{\partial x^2} \cdot \frac{b\,(p+n)}{(b+1)\,\delta p + bn_{n0} + p_{n0}} - \frac{p - p_{n0}}{\tau_p} -$$

$$- \frac{\partial p}{\partial x} \cdot \frac{jb\,(n_{n0} - p_{n0})}{q\,[(b+1)\,\delta p + bn_{n0} + p_{n0}]^2} + \left(\frac{\partial p}{\partial x}\right)^2 \frac{D_p\,(b-1)\,b\,(n_{n0} - p_{n0})}{[(b+1)\,\delta p + bn_{n0} + p_{n0}]^2}, \tag{1.15}$$

$$E = \frac{j}{q\mu_p\,[\delta p\,(b+1) + bn_{n0} + p_{n0}]} - \frac{kT}{q} \frac{\partial p}{\partial x} \frac{b-1}{(b+1)\,\delta p + bn_{n0} + p_{n0}}, \tag{1.16}$$

$$j_p = -qD_p \left[\frac{p\,(0,\,t)}{p\,(0,\,t) + n_{n0} - p_{n0}} + 1 \right] \frac{\partial p\,(x,\,t)}{\partial x}\Big|_{x=0}, \tag{1.17}$$

where $b = \mu_n / \mu_p$.

Although these transformations simplify the system (1.3)-(1.8), we still encounter considerable mathematical difficulties in the solution of the differential equation (1.15). Stafeev [16] has pointed out that at low injection levels ($p/n_{n0} \ll 1$) the third and fourth terms on the right-hand side of Eq. (1.15) are of the order of p/n_{n0} and at high injection levels ($p/n_{n0} \gg 1$) the same terms are of the order of n_{n0}/p, i.e., in both cases, these terms are small. Therefore, if we neglect these two terms we shall not commit a serious error in the calculation of the carrier distribution. Noting that at high injection levels $p \approx n \approx \delta p$, we obtain, using Eq. (1.15),

$$\frac{1}{D_p} \cdot \frac{\partial p}{\partial t} = \frac{\partial^2 p}{\partial x^2} - \frac{p - p_{no}}{l_p^2}, \qquad (1.18)$$

where $l_p = L_p \sqrt{2b/(b+1)}$ [$b = \mu_n/\mu_p$] and, since b ≈ 2 for germanium and silicon, the value of l_p is close to the diffusion length of the holes in the base.

Thus, at high and low injection levels the behavior of the holes in the diode base is described by a diffusion equation of the type given in Eq. (1.13) but with L_p replaced by a parameter of similar magnitude: $l_p = L_p \sqrt{2b/(b+1)}$.

However, we must remember that at high injection levels the hole lifetime may differ considerably from the lifetime at low injection levels and, consequently, the corresponding diffusion lengths may be different. Therefore, the identity of Eqs. (1.13) and (1.18) should be understood in the sense that these two equations have the same form and that the coefficient in front of $(p - p_{no})$ in Eq. (1.13) is equal to the diffusion length of holes, whereas in Eq. (1.18) this coefficient is close to the diffusion length if the value of L_p is measured at the appropriate injection level.

When τ_p = const irrespective of the injection level, the near-identity of Eqs. (1.13) and (1.18) means physically that the "pulling" of holes into the base by the electric field (which is produced when $p \gtrless n_{no}$) is slight.

It is worth mentioning that the assumption of the charge neutrality of a semiconductor, which is used as one basis of the transformation of the system (1.3)-(1.8), is not self-consistent. The form of the expression for the field intensity in the base, given by Eq. (1.16), shows that $\partial E/\partial x \neq 0$ and when this condition is substituted into Poisson's equation (1.8), we find that the electrical neutrality condition given by Eq. (1.10) is not obeyed.

However, the estimates in [17] show that at the maximum value of $\partial E/\partial x$ that has been observed experimentally in certain types of germanium diodes, the value of $|\delta_p - \delta_n|$ does not exceed 10^9 cm^{-3}. The equilibrium electron and hole densities in such diodes are, respectively, $n_{no} \approx 10^{14}$ cm^{-3} and $p_{no} \approx 10^{12}$ cm^{-3}. Thus, the difference $|\delta p - \delta n|$ is several orders of magnitude smaller than the equilibrium density of the minority carriers. This result is physically self-evident because, in the presence of a high density

of carriers, even a slight spatial separation produces very strong
fields, which tend to re-establish the neutrality, i.e., the difference
| $\delta p - \delta n$ | should remain small. When the excess carrier density
is increased, this difference naturally increases but the condition
(1.10) is, in practice, satisfied at least up to values of the hole density given by $p \lesssim (10-100)n_{n0}$ [17].

Thus, the assumption on which our transformation is based
represents effectively the quasineutrality condition $\delta p \approx \delta n$, which
implies the neglecting of the coordinate dependence of the electric
field intensity in Poisson's equation and the simultaneous use of
this dependence in Eq. (1.16) for E deduced from Eqs. (1.3) and (1.4).

In the majority of practical applications of diodes, particularly fast-response diodes (prepared from semiconductors of relatively low resistivities and high values of n_{n0}), the forward current densities are such that the condition $p < 100_{n0}$ is well satisfied.

When the current density is increased still further, the rise
of the density of injected holes slows down and this can be accounted
for only by a departure from the quasineutrality condition. At very
high current densities, the hole density near a p-n junction may
even decrease so that a large space charge is formed [18]. Under
these conditions, the transformations of the system (1.3)-(1.8) just
described are no longer valid.

The low-injection-level approximation, in which Eq. (1.13) is
valid, has been investigated most thoroughly and consistently. In
this case, the majority of transient processes can be described by
relatively simple expressions which can be checked experimentally.
These expressions are frequently valid also at higher injection levels, as condirmed by additional theoretical calculations or by experiments. Even when the exact form of a given formula is invalid
at an arbitrary injection level, such a formula still gives reliable
semiquantitative results..

The use of the full system of equations (1.3)-(1.8) is manda-
tory only when a diode is switched from the neutral to the forward
direction.

In view of this, we shall concentrate mainly on transient processes at low injection levels.

We must now estimate the values of the lifetime for which
the obtained equations are valid. It is known that the motion

Table 1.1

n_{n0}, cm^{-3}	l_D, cm	$\theta_M = \frac{\varepsilon}{4\pi\sigma}$,sec
$1 \cdot 10^{20}$	$3.96 \cdot 10^{-8}$	$0.63 \cdot 10^{-15}$
$1 \cdot 10^{17}$	$1.25 \cdot 10^{-6}$	$0.63 \cdot 10^{-12}$
$1 \cdot 10^{14}$	$3.96 \cdot 10^{-5}$	$0.63 \cdot 10^{-9}$
$1 \cdot 10^{11}$	$1.25 \cdot 10^{-3}$	$0.63 \cdot 10^{-6}$

of holes in a semiconductor obeys the diffusion equation only if the linear dimension, representing the hole density gradient (in our case, the diffusion length L_p), is considerably larger than the Debye screening length. The Debye screening length (or radius) is defined by the formula

$$l_D = \sqrt{\frac{\varepsilon kT}{4\pi q^2 n_{n0}}} . \qquad (1.19)$$

This condition is equivalent to the requirement that the lifetime of nonequilibrium carriers should exceed considerably the dielectric (Maxwellian) relaxation time θ_M. This relaxation time represents the time necessary for the re-establishment of charge neutrality in a semiconductor whenever this neutrality is disturbed in some way (for example, by the injection of holes at a p-n junction); this relaxation time is given by

$$\theta_M = \frac{\varepsilon \rho_n}{4\pi} ,$$

where ρ_n is the resistivity of the base. The values of the quantities l_D and θ_M are listed in Table 1.1 for n-type semiconductors used in the manufacture of semiconductor diodes.

We must mention that, even when special measures are taken, it is not possible to obtain $\tau_p < 10^{-6}$ sec in germanium or silicon with $n_{n0} \approx 10^{14}$ cm^{-3}, but when $n_{n0} = 10^{16}$ cm^{-3} the hole lifetime can be reduced to values of the order of 10^{-9}-10^{-10} sec. Thus, the condition $\tau_p \gg \theta_M$ is always obeyed and Eqs. (1.13) and (1.18) can be assumed to be valid at any given moment.

§3.   SOLUTION OF THE DIFFUSION EQUATION
(AT LOW INJECTION LEVELS)

3.1. Diode Model and Boundary Conditions

We shall consider the transient switching process in a diode at a low injection level and we shall ignore the drift current of the

minority carriers (holes), i.e., we shall solve the diffusion equation (1.13) on the assumption that in the circuit shown in Fig. 1.1
a forward current of density j_f flows for a very long time, and that
at a certain moment (from which we shall measure time) a reverse
voltage U_r is suddenly applied to the imput of the circuit.

To begin with, we must make additional assumptions about
this diode model and determine the initial and boundary conditions.

We shall assume that the change in the type of conduction in-
side a semiconductor crystal takes place suddenly at some value
of the coordinate x, which we shall regard as the origin. This assumption is satisfied by alloyed p-n junctions and, in the first approximation, by many diffused junctions.

The assumption about the sudden change from the p- to n-
type conductivity means that the injection efficiency of the contact,
γ, is close to unity and that the whole current in the p-n junction
plane consists of holes. It follows that the investigated transient
process is governed solely by the behavior of the carriers in the
base since a change in the diode circuit current does not alter at
all the carrier density in the p-type region. Therefore, we shall
consider only the base of a diode in which the minority carriers
are holes and the majority carriers are electrons.

Sah, Noyce, and Shockley [104] have established that the val-
ue of γ can be considerably greater than unity when $\rho_p \ll \rho_n$ because of strong recombination of holes in the space-charge region
near a p-n junction. This effect may be important in semiconductors with a wide forbidden band, a low resistivity, and short lifetimes. In practice, recombination in a p-n junction governs the current-voltage characteristics of the majority of silicon diodes at
low forward currents and the characteristics of germanium diodes
at low temperatures. Recombination in a p-n junction can also be
important at room temperature in the case of some fast-response
germanium diodes with low base resistivities and short nonequilibrium carrier lifetimes [159].

When the forward current is increased, the electron-hole recombination current also increases but more slowly than the whole
diffusion current. Thus, at sufficiently high values of j_f, the assumption that $\gamma = 1$ is valid. At lower forward currents when $\gamma < 1$,
the hole current fraction is γj_f and, as in the case $\gamma = 1$, no changes

take place in the carrier density in the p–type region because electrons flowing from the n–type region recombine in the space–charge
layer. Consequently, even when $\gamma < 1$ (provided this is due to recombination in a p–n junction), we need consider only those processes which take place in the diode base.

We shall continue our analysis on the assumption that $\gamma = 1$,
bearing in mind that, if this condition is not satisfied, all the formulas remain valid for the hole component of the total current.

When a current of density j flows through a p–n junction and
there is no electric field in the base, the boundary condition for the
p–n junction plane can be obtained easily from Eqs. (1.3) and (1.5):

$$\left(\frac{\partial p}{\partial x}\right)_{x=0} = -\frac{j}{qD_p}. \tag{1.20}$$

We shall also assume that the carrier densities in the base are nev-
er high enough to disobey the Boltzmann statistics. In the case of
germanium and silicon near room temperature, this assumption is
valid to carrier densities of at least 10^{18} cm^{-3}. The use of the
Boltzman approximation allows us to relate the hole density at the
boundary of the space–charge region (on the base side) p_1, which we
shall call the impressed density, to the voltage u_{p-n} across a p–n
junction by the following simple expression:

$$p_1 = p_{n0} \exp\left(\frac{qu_{p-n}}{kT}\right). \tag{1.21}$$

It follows that if a sufficiently large forward voltage ($u_{p-n} \gg kT/q$),
is applied to a p–n junction, we find that $p_1 \gg p_{n0}$; when a reverse
voltage is applied we can assume with a high degree of accurcy that

$$p(0,\ t) = 0. \tag{1.22}$$

As we move away to infinity from a p–n junction, the density
of the excess holes decreases to zero because of recombination, so
that the condition at infinity has the form

$$p(x,\ t) \to p_{n0} \quad \text{when} \quad x \to \infty. \tag{1.23}$$

Thus, if we know the time dependence of the voltage drop across
the p–n junction, $u_{p-n}(t)$, we are dealing with a boundary problem

of the first kind (having determined the impressed hole density);
if we know the time dependence of the density of the current flowing
through the p-n junction, j(t), we are dealing with a boundary problem of the second kind (the derivative of the hole density with respect to the position coordinate is known at the boundary).

When a forward current of density j_f flows through a diode
for an infinitely long time, the distribution of the holes in the base
is given by the solution of the diffusion equation (1.13) in the steady-
state case ($\partial_p/\partial_t = 0$) with the boundary condition given by Eq. (1.20)
or Eq. (1.21). This solution has been obtained by Shockley [2] and
is of the form

$$p(x) = (p_1 - p_{n0}) \exp\left(-\frac{x}{L_p}\right) + p_{n0}. \qquad (1.24)$$

A simple relationship between the impressed hole density and the
forward current density can be obtained from Eqs. (1.20) and (1.24):

$$p_1 = \frac{j_f \, L_p}{q D_p} + p_{n0} = \frac{j_f \, \tau_p}{q L_p} + p_{n0} \qquad (1.25)$$

or, since under practical switching conditions in diodes, we have
$p_1 \gg p_{n0}$,

$$p_1 \simeq \frac{j_f \, L_p}{q D_p}. \qquad (1.26)$$

Thus, when the forward bias is increased, the impressed hole
density increases exponentially with increasing voltage across the
p-n junction, and linearly with increasing current through the junction.

The distribution of holes given by Eq. (1.24) is the required
initial condition for the solution of the problem of the switching of
a diode from the forward to the reverse condition.

3.2. Methods for Solving the Diffusion Equation

We shall now find the integral of the diffusion equation for
both boundary conditions, i.e., we shall consider two problems:

1) a forward current of density j_f flows for an infinitely long
time through a diode but at a moment t = 0 a reverse voltage U_r

is applied instantaneously to the p-n junction: this is represented by the boundary condition (1.22);

2) a forward current again flows for an infinitely long time and the diode is switched to the reverse polarity at a moment $t = 0$ in such a way that a constant reverse current of density j_0 begins to flow through the diode, as given by the boundary condition (1.20).

To simplify further mathematical analysis, we shall introduce dimensionless time and space coordinates:

$$\mathcal{J} = \frac{t}{\tau_p} \quad \text{and} \quad X = \frac{x}{L_p}. \qquad (1.27)$$

When these dimensionless variables are used, Eq. (1.13), the boundary conditions (1.20) and (1.22), and the initial condition (1.24) become

$$\frac{\partial p}{\partial \mathcal{J}} = \frac{\partial^2 p}{\partial X^2} - (p - p_{n0}), \qquad (1.28)$$

$$\left(\frac{\partial p}{\partial X}\right)_{X=0} = -\frac{j_0 L_p}{q D_p} = -\frac{j_0 \tau_p}{q L_p}, \qquad (1.29)$$

$$p(0, \mathcal{J}) = 0, \qquad (1.30)$$

$$p(X, 0) = \frac{j_f L_p}{q D_p} e^{-X} + p_{n0}, \qquad (1.31)$$

where $p = p(X, \mathcal{J})$.

We shall now consider the first boundary problem. Making the substitution

$$p - p_{n0} = \varphi(X, \mathcal{J}) e^{-\mathcal{J}}, \qquad (1.32)$$

we reduce the diffusion equation to its simplest form

$$\frac{\partial \varphi}{\partial \mathcal{J}} = \frac{\partial^2 \varphi}{\partial X^2}. \qquad (1.33)$$

We shall also rewrite the conditions (1.23), (1.30), and (1.31):

$$\varphi(0, \mathcal{J}) = -p_{n0}e^{\mathcal{J}}, \tag{1.34}$$

$$\varphi(\infty, \mathcal{J}) = 0, \tag{1.35}$$

$$\varphi(X, 0) = (p_1 - p_{n0})e^{-X}. \tag{1.36}$$

The solution of the problem can be obtained in the form of the sum [19]:

$$\varphi(X, \mathcal{J}) = \varphi_1(X, \mathcal{J}) + \varphi_2(X, \mathcal{J}), \tag{1.37}$$

where $\varphi_1(X,\mathcal{J})$ represents the influence of the initial conditions only and a $\varphi_2(X, \mathcal{J})$ represents the influence of the boundary condition only.

These functions can be found by solving Eq. (1.33) and making the solutions satisfy the conditions:

I. $\varphi_1(X, 0) = \varphi(X, 0)$, $\varphi_1(0, \mathcal{J}) = 0$, $\tag{1.38}$

II. $\varphi_2(X, 0) = 0$, $\varphi_2(0, \mathcal{J}) = \varphi(0, \mathcal{J})$. $\tag{1.39}$

It is obvious that the sum of these functions satisfies the boundary and initial conditions given by Eqs. (1.34) and (1.36).

Thus solutions for $\varphi_1(X, \mathcal{J})$ and $\varphi_2(X, \mathcal{J})$, obtained using the source function for a semi-infinite line are already known (cf., for example, [19, 20]):

$$\varphi_1(X, \mathcal{J}) = \frac{1}{2\sqrt{\pi\mathcal{J}}} \int_0^\infty \varphi_1(\xi)\left\{ \exp\left[-\frac{(X-\xi)^2}{4\mathcal{J}}\right] - \exp\left[-\frac{(X+\xi)^2}{4\mathcal{J}}\right]\right\} d\xi, \tag{1.40}$$

$$\varphi_2(X, \mathcal{J}) = \frac{2}{\sqrt{\pi}} \int_{\frac{X}{2\sqrt{\mathcal{J}}}}^\infty \varphi_2\left(\mathcal{J} - \frac{X^2}{4\xi^2}\right)\exp(-\xi^2)\,d\xi. \tag{1.41}$$

The integration of Eqs. (1.40) and (1.41) and the inversion back to the function $p(X, \mathcal{J})$ give the required solution

$$p(X, \mathcal{J}) = p_{n0}(1 - e^{-X}) + \frac{p_1}{2}\left[e^{-X}\operatorname{erfc}\left(\sqrt{\mathcal{J}} - \frac{X}{2\sqrt{\mathcal{J}}}\right) - \right.$$

$$\left. - e^X\operatorname{erfc}\left(\sqrt{\mathcal{J}} + \frac{X}{2\sqrt{\mathcal{J}}}\right)\right]^* \tag{1.42}$$

The most widely used method in solving the diffusion equation for various transient processes in diodes is the operational calculus method. To demonstrate the advantages of this method, we shall re-derive Eq. (1.42).

The application of the operator method to second-order differential equations with two variables is based on the two-dimensional Laplace-Carson transformation. The basis of this transformation is that each function of the arguments X and $\mathcal{J}$, used in the solution of the differential equation (1.33), is matched by some functions of the variables s_1 and s_2, which are defined as follows

$$\varphi(X, \mathcal{J}) \doteqdot F(s_1, s_2) \equiv s_1 \cdot s_2 \int\limits_0^\infty \int\limits_0^\infty \exp(-s_1 X - s_2 \mathcal{J})\varphi(X, \mathcal{J})\,dX\,d\mathcal{J}, \tag{1.43}$$

$$\varphi(X, 0) \doteqdot A_1(s_1) \equiv s_1 \int\limits_0^\infty \exp(-s_1 X)\varphi(X)\,dX, \tag{1.44}$$

$$\varphi(0, \mathcal{J}) \doteqdot A_2(s_2), \tag{1.45}$$

$$\left(\frac{\partial\varphi}{\partial X}\right)_{X=0} \doteqdot A_3(s_2), \tag{1.46}$$

where the expressions for the function $A_2(s_2)$ and $A_3(s_2)$ are fully analogous to Eq. (1.44). The functions of s_1 and s_2 on the right-

---

* The generally accepted abbreviation for complementary error function is

$$\operatorname{erfc} Z \equiv 1 - \operatorname{erf} Z = 1 - \frac{2}{\sqrt{\pi}}\int\limits_0^Z e^{-y^2}\,dy.$$

hand side of Eqs. (1.43)-(1.46) are called the transforms and the functions on the left-hand side are called the originals. When the originals are replaced by the transforms, the differential equation (1.83) transforms to an algebraic equation which can be solved for a semi-infinite region [21]

$$F(s_1, s_2) = \frac{-s_2 A_1(s_1) + s_1 A_3(s_2) + s_1^2 A_2(s_2)}{s_1^2 - s_2}. \qquad (1.47)$$

The transforms of the initial and boundary conditions are not independent but related by the "compatibility condition":

$$s A_1(\sqrt{s}) - \sqrt{s}\, A_3(s) - s A_2(s) = 0, \quad s = s_1 \quad \text{or} \quad s = s_2, \qquad (1.48)$$

from which it follows that only two of these functions can be selected arbitrarily. In the boundary problem which we are considering, we know $A_1(s_1)$ and $A_2(s_2)$ and, therefore, we can find $A_3(s_2)$ from Eq. (1.48); substituting the latter function in Eq. (1.47), we obtain (after suitable algebraic transformations) the function $F(s_1, s_2)$ in the following form

$$F(s_1, s_2) = \frac{s_1 s_2}{(s_1 + 1)(\sqrt{s_2} + 1)(s_1 + \sqrt{s_2})} p_1 - \frac{s_1 s_2}{(s_1 + 1)(s_2 - 1)} p_{n0}. \qquad (1.49)$$

We can easily show that the coefficient in front of p_1 can be represented thus:

$$\frac{s_1 s_2}{(s_1 + 1)(s_2 - 1)} - \frac{1}{2} \frac{s_1 \sqrt{s_2}}{(\sqrt{s_2} - 1)(\sqrt{s_2} + s_1)} - \frac{1}{2} \frac{s_1 \sqrt{s_2}}{(\sqrt{s_2} + 1)(\sqrt{s_2} + s_1)}.$$

Using the tables given in [21] to find the originals of the function s_1 and s_2 and performing the inversion from $\psi(X, \mathcal{T})$ to $p(X, \mathcal{T})$, we again obtain the post-switching distribution of the hole density in the base in the form of Eq. (1.42).

One of the advantages of the operator method is that Eq. (1.48) can be used to find easily either $A_3(s_2)$ or $A_2(s_2)$. The inversion back to the originals makes it possible immediately to obtain the time dependence of the density of the current through the p-n junction or of the density of the holes near the junction without solving completely Eq. (1.33). A knowledge of these two dependences is frequently sufficient for the description of a transient process.

Another advantage of the operator method is the relative ease with which a given transform can be determined from Eq. (1.47) or (1.48) and the simplicity of finding the original from the transform using tables, some of which are given in [21–25].

The use of the two-dimensional Laplace-Carson transformation simplifies the problem of finding the solution of Eq. (1.33) for other initial and boundary conditions.

Thus, solving the second boundary problem, when the density of the reverse current after switching is constant and equal to j_0, we obtain

$$p(X, \mathcal{J}) = p_1 \left\{ e^{-X} - \frac{j_f + j_0}{2j_f} \left[e^{-X} \operatorname{erfc} \left(\frac{X}{2\sqrt{\mathcal{J}}} - \sqrt{\mathcal{J}} \right) \right.\right.$$

$$\left.\left. - e^X \operatorname{erfc} \left(\frac{X}{2\sqrt{\mathcal{J}}} + \sqrt{\mathcal{J}} \right) \right] \right\}. \qquad (1.50)$$

For brevity, we shall introduce the functions

$$\left. \begin{aligned} \psi_1(X, \mathcal{J}) &= \frac{1}{2} \left[e^{-X} \operatorname{erfc} \left(\frac{X}{2\sqrt{\mathcal{J}}} - \sqrt{\mathcal{J}} \right) - \right.\\ &\qquad \left. - e^X \operatorname{erfc} \left(\frac{X}{2\sqrt{\mathcal{J}}} + \sqrt{\mathcal{J}} \right) \right],\\ \psi_2(X, \mathcal{J}) &= \frac{1}{2} \left[e^{-X} \operatorname{erfc} \left(\sqrt{\mathcal{J}} - \frac{X}{2\sqrt{\mathcal{J}}} \right) - \right.\\ &\qquad \left. - e^X \operatorname{erfc} \left(\sqrt{\mathcal{J}} + \frac{X}{2\sqrt{\mathcal{J}}} \right) \right],\\ \psi_3(X, \mathcal{J}) &= \frac{1}{2} \left[e^{-X} \operatorname{erfc} \left(\frac{X}{2\sqrt{\mathcal{J}}} - \sqrt{\mathcal{J}} \right) + \right.\\ &\qquad \left. + e^X \operatorname{erfc} \left(\frac{X}{2\sqrt{\mathcal{J}}} + \sqrt{\mathcal{J}} \right) \right],\\ \psi_4(X, \mathcal{J}) &= \frac{1}{2} \left[e^{-X} \operatorname{erfc} \left(\sqrt{\mathcal{J}} - \frac{X}{2\sqrt{\mathcal{J}}} \right) + \right.\\ &\qquad \left. + e^X \operatorname{erfc} \left(\sqrt{\mathcal{J}} + \frac{X}{2\sqrt{\mathcal{J}}} \right) \right]. \end{aligned} \right\} \qquad (1.51)$$

We can easily see that these functions are related by the following simple expressions:

$$\psi_1(X, \mathcal{J}) + \psi_4(X, \mathcal{J}) = \psi_2(X, \mathcal{J}) + \psi_3(X, \mathcal{J}) = e^{-X}. \qquad (1.52)$$

Using these contractions, we can write Eq. (1.42) in the form

$$p(X, \mathcal{J}) = p_{n0}(1 - e^{-X}) + p_1\psi_2(X, \mathcal{J}) \qquad (1.53)$$

and, correspondingly, Eq. (1.50) becomes

$$p(X, \mathcal{J}) = p_1[e^{-X} - (1 + B)\psi_1(X, \mathcal{J})], \qquad (1.54)$$

where $B = j_0/j_f$.

A distribution of the type given by Eq. (1.42) was first derived by Hebb and again by Pell [11], while Eq. (1.50) has been obtained by Kingston [8] and in an explicit form by Lax and Neustadter [9].

Figure 1.2 gives solutions of Eqs. (1.42) and (1.50) for four different moments of time, when the condition $p_{n0} \ll p_1$ is satisfied.

The change in the initial distribution of holes at the moment $\mathcal{J} = 0$ begins near the p-n junction and spreads into the base later. The base loses the excess holes most rapidly when a reverse voltage pulse is applied to the p-n junction and the hole density changes suddenly, in accordance with the assumed boundary condition (1.22),

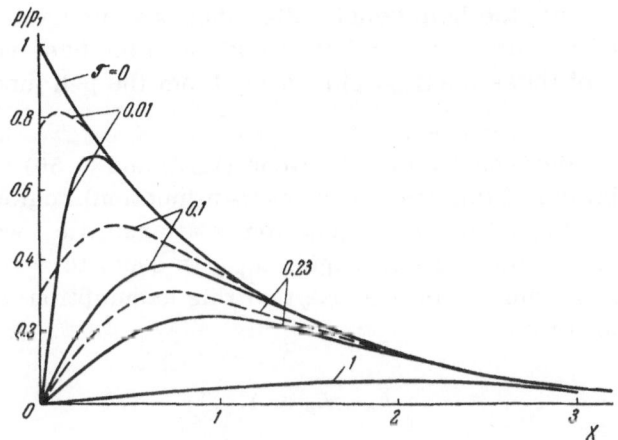

Fig. 1.2. Distributions of the relative hole densities p/p_1 at various times after switching. The continuous curves represent switching without a limiting resistance; the dashed curves represent switching for $j_0/j_f = 1$.

from p_1 to zero. Physically, this means that a negative bias across the p-n junction makes it a perfect absorber of holes: any hole approaching the p-n junction due to diffusion is immediately pulled into the junction and swept into the p-type region. From physical considerations it is obvious that since holes are lost from the base only by the diffusion or recombination processes, the rate of dispersal should be independent of the value of the reverse voltage since this voltage is always concentrated in the space-charge region. The form of the distribution given by Eq. (1.42) confirms this conclusion.

It is worth noting that in the interior of the base the distribution p(x) retains its initial form for some (shorter or longer) time after switching. This is because the loss of holes due to recombination, which takes place continuously at each point in the base, is made up by holes diffusing in the direction of decreasing hole density (from left to right in Fig. 1. 2). It follows naturally from Eqs. (1.6), (1.20), and (1.24) that far from the p-n junction the loss and gain of holes are exactly equal. The disturbance of the steady-state distribution spreads slowly into the interior of the base.

If, after the application of a reverse voltage step, the current through the p-n junction is limited, the number of holes near the junction decreases to zero in a finite time.

In all cases, the hole density distributions during dispersal are described by curves which have maxima in the base region; the positions of these maxima shift away from the p-n junction with time.

We have derived the distributions (1.42) and (1.50) neglecting the redistribution of the space charge (p-n junction) region when the external voltage across the junction is altered, i.e., we have assumed that the width of this region h_{p-n} is equal to zero. It is evident from the curves in Fig. 1.2 that this assumption is equivalent to the condition

$$h_{p-n}/L_p \ll 1.$$

For the majority of germanium and silicon planar diodes of the alloyed and diffused type, this condition is well satisfied provided the reverse voltage is not too high. The neglecting of the nonstationary processes taking place in the space-charge region means that we are ignoring the influence of the barrier capacitance on the duration of the transient process.

3.3. Establishment of a Hole Distribution

Under a Forward Bias

In many cases, it is incorrect to assume that the forward current before switching has been flowing through the diode for an infinitely long time and the distribution of holes in the base reached its steady state. In the cases of a short forward–current pulse, we cannot use Eq. (1.24) as the initial equation for the solution of Eq. (1.13); we must consider instead the process of the establishment of a steady–state distribution after the application of a forward current pulse [26, 27].

A sudden increase of the forward current density from zero to j_f results immediately in the establishment of a hole density gradient near the p–n junction and this gradient is given by Eq. (1.20). Before the application of this forward pulse, the density of holes in the diode base is

$$p(X, 0) = p_{n0}, \quad t = 0.$$

The solution of the diffusion equation (1.13) with this initial distribution and with the boundary condition given by Eq. (1.20) has the form

$$p = p_1 \frac{1}{2} \left\{ e^{-X} \operatorname{erfc}\left(\frac{X}{2\sqrt{\mathcal{J}_f}} - \sqrt{\mathcal{J}_f}\right) - \right.$$
$$\left. - e^X \operatorname{erfc}\left(\frac{X}{2\sqrt{\mathcal{J}_f}} + \sqrt{\mathcal{J}_f}\right) \right\} + p_{n0} = p_1 \psi_1(X, \mathcal{J}_f) + p_{n0}. \quad (1.55)$$

Here, $\mathcal{J}_f = t_f/\tau_p$, where t_f is the duration of the forward current pulse. The dependences $p(X)/p_1$ for various fixed values of $\mathcal{J}_f$ are shown in Fig. 1.3. Examination of the curves in that figure shows [26] that, if we assume $p_{n0} = 0$, we find that the distribution of the excess holes in the base for $\mathcal{J}_f \ll 1$ can be described approximately by an exponential dependence of the type

$$p(X, \mathcal{J}_f) = p_1(\mathcal{J}_f) \exp\left[-x/l(\mathcal{J}_f)\right]. \quad (1.56)$$

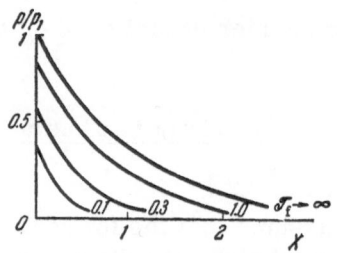

Fig. 1.3. Distributions of the relative hole densities in the base of a
diode for various durations of a for
ward current pulse.

We shall now determine the functions $p_1(\mathcal{J}_f)$ and $l(\mathcal{J}_f)$. To do this, we shall require the approximate and exact distributions to have the same value of the hole density at the p-n junction and the same total number of excess holes in the whole base. We shall use the following asymptotic expansions of the error function in the form of a series for small and large values of the argument:

$$\left.\begin{aligned}
\operatorname{erf} z &\simeq \frac{2}{\sqrt{\pi}}\left(z-\frac{z^3}{1!3}+\frac{z^5}{2!5}-\cdots\right), \\
\operatorname{erfc} z &\simeq \frac{1}{\sqrt{\pi}}\frac{e^{-z^2}}{z}\left[1-\frac{1}{2z^2}+\frac{3\cdot4}{(2z)^4}-\frac{4\cdot5\cdot6}{(2z)^6}+\cdots\right]
\end{aligned}\right\}. \qquad (1.57)$$

From the general solution given by Eq. (1.55), we obtain, for $X = 0$,

$$p_1(\mathcal{J}_f) = p_1 \cdot \operatorname{erf}\sqrt{\mathcal{J}_f} \simeq p_1\frac{2}{\sqrt{\pi}}\sqrt{\mathcal{J}_f}. \qquad (1.58)$$

Integrating Eq. (1.13) over the volume of the base and noting that

$$Sq\int_0^\infty p\,dx = Q_{st}$$ is the total charge of holes stored in the base, we

obtain the following equation for the charge accumulation process:

$$Q_{st}(\mathcal{J}_f) = Q_{st}(\mathcal{J}_f \to \infty)(1 - e^{-\mathcal{J}_f}), \qquad (1.59)$$

where $Q_{st}(\mathcal{J}_f \to \infty) = Sqp_1L_p$. When $\mathcal{J}_f \ll 1$, we obtain

$$Q_{st}(\mathcal{J}_f) = \mathcal{J}_f\,Q_{st}(\mathcal{J}_f \to \infty) = Sqp_1L_p\mathcal{J}_f; \qquad (1.60)$$

integration of Eq. (1.56), multiplied by qS, over the volume of the base gives

$$Q_{st}(\mathcal{J}_f) = p_1(\mathcal{J}_f)\, l(\mathcal{J}_f)\, qS. \tag{1.61}$$

Equating Eqs. (1.60) and (1.61) we obtain

$$l(\mathcal{J}_f) = L_p \frac{\mathcal{J}_f}{\operatorname{erf}\sqrt{\mathcal{J}_f}} \simeq L_p \frac{\sqrt{\pi}}{2} \sqrt{\mathcal{J}_f} \; . \tag{1.62}$$

Thus, the hole density distribution in the base at various times after the application of a forward current pulse is described approximately by the expression

$$p(X, \mathcal{J}) = p_1 \operatorname{erf}\sqrt{\mathcal{J}_f}\, \exp\left(-\frac{\operatorname{erf}\sqrt{\mathcal{J}_f}}{\mathcal{J}_f}\, X\right) \simeq$$

$$\simeq p_1 \frac{2}{\sqrt{\pi}} \sqrt{\mathcal{J}_f}\, \exp\left(-\frac{2}{\sqrt{\pi}\,\mathcal{J}_f}\, X\right), \tag{1.63}$$

where p_1 is defined, as before, by Eq. (1.25). Expressions similar to Eq. (1.63) have been reported also in [26, 27, 28]. Numerical estimates obtained for $\mathcal{J}_f = 0.1$ show that the expressions (1.63) and (1.55) differ by not more than 3% for a given value of X.

Assuming approximately that $2/\sqrt{\pi} \approx 1$, we find that the distribution of holes in the base for short forward current pulses ($t \ll \tau_p$) is the same as in the steady-state case if the impressed density of holes near the p-n junction (p_1) and the diffusion length (L_p) in the expression for the steady-state case are both reduced by a factor of $1/\sqrt{\mathcal{J}_f}$.

When a forward bias is applied to a diode and the conditions in the circuit are such that the external voltage is concentrated instantaneously in the p-n junction, the impressed density of holes increases suddenly to p_1, i.e., the relationship (1.21) applies even at $t = 0$. We shall now integrate the diffusion equation (1.13) for the zero initial condition and for the boundary condition given by

$$p(0, t) = p_1. \tag{1.64}$$

The solution is obtained in the same way as in the case of a constant forward current and this solution is

$$p(X, \mathcal{J}_f) = p_1 \psi_3(X, \mathcal{J}_f). \tag{1.65}$$

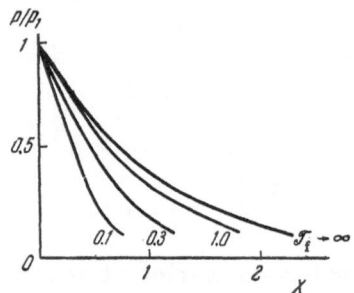

Fig. 1.4. Distributions of the relative hole densities in the base of a diode for various durations of a forward voltage pulse.

The graphs of the function (1.65) are given in Fig. 1.4 for various values of $\mathcal{T}_f$. Comparison of Figs. 1.3 and 1.4 shows clearly the characteristic differences between the two conditions: the constancy of either the density gradient or of the hole density itself at the p-n junction during the transient process, provided j_f = const or u_{p-n} = const in these two cases.

The problem of switching a diode from a nonstationary forward-current state can be solved using as the initial condition the distributions given by Eqs. (1.55), (1.63), or (1.65). From the form of these expressions and from Figs. 1.3 and 1.4, it is clear that a forward current pulse should be regarded as "short" when its duration is comparable with or shorter than the hole lifetime.

Chapter II

Switching in a Planar Diode

We shall now consider the transient processes which accompany various forms of the switching of a diode from the forward to the reverse direction. As in Chap. I, we shall use a model diode with a planar p-n junction and a semi-infinite n-type base; we shall simply call such a structure a planar diode.

§4. TRANSIENT PROCESSES WITHOUT A LIMITING RESISTANCE IN THE DIODE CIRCUIT

4.1. General Solution

We shall find the transfer function after the establishment of a steady forward current in the diode in the circuit shown in Fig. 1.1 but with $R_l = 0$, i.e., we shall assume that a reverse voltage is applied instantaneously to the p-n junction at some moment $t = 0$ ($\mathcal{T} = 0$). We shall show that such an assumption is often the best approximation to the experimental conditions. The condition $R_l = 0$ means that Eq. (1.13) should be solved using the boundary condition given by Eq. (1.22).

Substituting Eq. (1.42) into the condition (1.20) and differentiating, we obtain the following expression for the density of the transient reverse current through a p-n junction:

$$ j(\mathcal{T}) = -\frac{qD_p p_{n0}}{L_p} - \frac{qD_p p_1}{L_p}\left(\frac{e^{-\mathcal{T}}}{\sqrt{\pi\mathcal{T}}} - \operatorname{erfc}\sqrt{\mathcal{T}}\right). \qquad (2.1) $$

Using Eq. (1.25) and the equation

$$ -\frac{qD_p p_{n0}}{L_p} = j_s \qquad (2.2) $$

which is simply the saturation density of the current flowing through the p-n junction,* we find that

$$\frac{j+j_s}{j_f+j_s} = -\left(\frac{e^{-\mathcal{J}}}{\sqrt{\pi \mathcal{J}}} - \text{erfc } \sqrt{\mathcal{J}}\right). \tag{2.3}$$

Usually, $j_f \gg j_s$ and the dependence $j(t)$ is most interesting when $j(t) \gg j_s$; therefore, assuming that $j_s = 0$ in Eq. (2.3), we obtain

$$j(t) = -j_f\left(\frac{e^{-\mathcal{J}}}{\sqrt{\pi \mathcal{J}}} - \text{erfc } \sqrt{\mathcal{J}}\right). \tag{2.4}$$

The minus sign in from of j_f means that $j(t)$ is directed antiparallel to j_f, i.e., it is the reverse current through the diode. We shall always consider the absolute value of the reverse current and, therefore, we shall omit the minus sign in similar formulas. We shall also neglect the saturation current because it is small compared with the instantaneous values of the reverse current transient. This is equivalent to the assumption that $p_{n0} = 0$ in Eqs. (1.24), (1.42), and (1.55) and in similar hole density distributions.

Expanding the exponential function and the error function of Eq. (1.57) in the form of a series and retaining only the terms of the first order of smallness for small values of $\mathcal{J}$ and the terms of the second order of smallness for large values of $\mathcal{J}$, we obtain

$$j(t) \simeq j_f \frac{1}{\sqrt{\pi \mathcal{J}}}, \quad \mathcal{J} \ll 1, \tag{2.5}$$

$$j(t) \simeq j_f \frac{e^{-\mathcal{J}}}{\sqrt{\pi \mathcal{J}^3}}, \quad \mathcal{J} \gg 1. \tag{2.6}$$

Curves showing the exact solution of Eq. (2.4) and its asymptotic approximations are present in Fig. 2.1. We can see that, to within a factor of 2, the approximation given by Eq. (2.5) is valid when $\mathcal{J} < 0.2$ and the approximation given by Eq. (2.6) is valid when $\mathcal{J} > 1$.

---

* The expression (2.2) can be obtained directly from Eq. (1.25) by replacing j_f with j_s and utilizing the condition $p_1 = 0$ for the reverse bias.

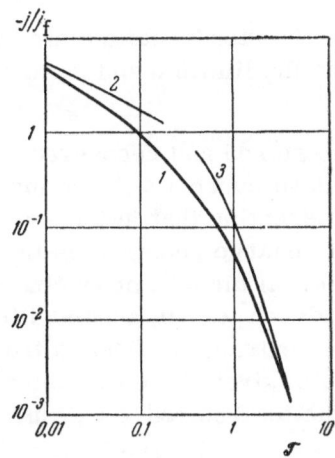

Fig. 2.1. Time dependence of the transient reverse current through a a planar diode for $R_l = 0$: 1) the exact solution given by Eq. (2.4); 2), 3) the asymptotic expressions given by Eqs. (2.5) and (2.6), respectively.

Time dependences of the reverse transient current of the type given by Eq. (2.4) have been obtained and analyzed in detail by many workers [8, 9, 11, 29, 30].

An important characteristic of the transient process is the value of the total charge flowing, after switching, in the external circuit; this charge is equal to the charge of excess holes which have reached the p-type region from the base. This charge is known as the recovered charge Q_{rec} and represents some fraction of the total charge of excess holes accumulated in the base of a diode during the flow of the forward current. Rediker and Sawyer [31] were the first to use the "stored charge" concept in the description of the response of diodes of various types.

The value of Q_{rec} is found by integrating the reverse transient current with respect to time:

$$Q_{rec} = S \int_0^\infty j(t)\, dt. \tag{2.7}$$

The substitution of the expression (2.4) into Eq. (2.7) gives

$$Q_{rec} = \frac{i_f \tau_p}{2} = \frac{Q_{st}}{2}, \tag{2.8}$$

where i_f is the forward current of the diode and Q_{st} is the stored hole charge. The second equality in Eq. (2.8) is obtained, in an elementary manner, by the integration of the steady-state distribution of holes, given by Eq. (1.24), between the limits 0 and ∞, neglecting the value of p_{n0}.

Analysis of Eq. (2.8) shows that the stored and recovered charges are both proportional to the forward current and are independent of the p-n junction area. It is interesting that in the switching process half of the stored excess hole charge recombines in the base and the other half flows into the external circuit, producing a transient reverse current. Since the condition $R_l = 0$, assumed in the calculation, ensures the most efficient extraction of holes from the base, the relationship given by Eq. (2.8) gives the maximum value of Q_{rec} considered as a fraction of the total stored charge under any switching conditions.

4.2. Refinement of the Boundary Conditions

It follows from Eqs. (2.4) and (2.5) that when $t \rightarrow 0$, the time dependence of the current density becomes $j(t) \rightarrow \infty$. This is the consequence of our assumption of an instantaneous change in the hole density at the p-n junction from p_1 to 0, on which the derivation of Eqs. (1.42) and (2.4) is based.

Obviously, an infinitely large current cannot flow through a p-n junction. In all the experiments on diodes it has been found that, although the reverse current may rise suddenly to very large values, these values are nevertheless finite.

In order to remove this contradiction between the theory and experiment, several attempts have been made to refine the boundary condition (1.22). Thus, Grinberg and Avak'yants [32] have analyzed the transient process on the assumption that $p(X = 0) = 0$ is not established instantaneously after the application of a reverse voltage to the diode. Some delay in the change of the hole density corresponding to the boundary condition is due to the fact that the charging of the barrier capacitance of the p-n junction requires a finite time, given by $t_C \approx R_l C_d$, where R_l is the load resistance in the diode circuit and C_d is the barrier capacitance of the diode which is assumed to be independent of the bias across the diode. Assuming that, after the switching, the voltage rise across this capacitance is described by a linear expression (which corresponds to an expansion of an exponential function as a series for small

values of t), Grinberg and Avak'yants have solved Eq. (1.13) and
found that the solution has a maximum in the time dependence of
the reverse current j = j(t). However, the calculations of Grinberg
and Avak'yants [32] are not completely satisfactory for two reasons.
First, the assumption that $t_C \neq 0$ does not agree with the assumption that $R_l = 0$ made in the derivation of Eq. (2.4). Secondly, if we
assume that $R_l \neq 0$, it follows that when a voltage is applied to the
p–n junction we must take into account not only the charging of the
barrier capacitance but also the voltage drop across the resistance
R_l during the flow of the transient reverse current through it (cf. § 5).

The inaccuracy of the calculations is confirmed by comparison with the results of an experimental investigation of diodes reported in [33] and discussed in [32]. To make the theoretical and
experimental results agree, it is necessary to assume that $R_l C_d =$
$2.66 \cdot 10^{-6}$ sec, which is at least 2-3 orders of magnitude larger
than the highest possible value of this quantity. The characteristics
reported in [33] were obtained for diodes with small-area p–n junctions, which were prepared by welding a thin gold wire to a germanium crystal. A typical value of the barrier capacitance for such
diodes does not exceed a few picofarads and the load resistance in
the switching processes is 1-2 k Ω.

Another attempt to remove the infinitely large reverse current from the theory has been made by Baranov and Bekbulatov [34],
who have used a time-dependent boundary condition for the p–n junction. They have assumed that, after the switching, the hole density
at the p–n junction decreases gradually in accordance with the law:

$$p(0, t) = p_1 e^{-\alpha t}, \qquad (2.9)$$

where $\alpha = \dfrac{qU_r}{kT} \cdot \dfrac{D_p}{L_p h_{p-n}} \cdot$ C for thick-base diodes. Using Eq. (2.9) we
can obtain an expression for the calculation of the reverse current
during switching.

However, it is necessary to point out that the condition (2.9)
is introduced in a formal manner without relation to the physical
processes taking place during the switching; therefore, its validity
is not self-evident. Moreover, the use of Eq. (2.9) removes only
the theoretically infinite values of the reverse current but not the
possibility of extremely large reverse currents. Let us consider
the following example: $U_r = 10$ V, $D_p = 44$ cm^2/sec, $L_p \approx 0.02$ cm

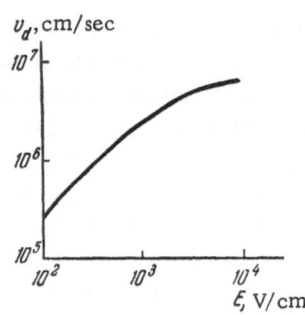

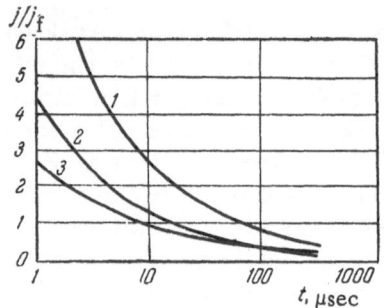

Fig. 2.2. Dependence of
the drift velocity of carriers in germanium on the
electric field intensity.

Fig. 2.3. Decay of the transient reverse current in diodes with different hole transit times through the p-n
junction for $v_d = \infty$ (1), $v_d = 10^6$ cm/sec
(2), and $v_d = 10^5$ cm/sec (3).

(i.e., $\tau_p = 10^{-5}$ sec), and $h_{p-n} = 2 \cdot 10^{-4}$ cm. In this case, $\alpha = 4 \cdot 10^9$
sec^{-1} and the hole density near the p-n junction decreases, in a time
interval of $(0.5-1) \cdot 10^{-9}$ sec, to a value much smaller than p_1, i.e.,
during this time interval after switching the reverse current
can be described by Eq. (2.4). It follows from this equation that for
$\mathcal{J} = (0.5-1) \cdot 10^{-4}$ the reverse current rises to a peak, which is 50–
100 times larger than the forward current.

Another attempt to refine Eq. (2.2) has been made by Scott
[35]. An infinitely large reverse current peak presupposes an infinitely high velocity of holes in the p-n junction region. However,
it is known that when the field intensity is increased, the carrier
mobility decraeses and tends to a constant value. A typical dependence of the drift velocity of carriers v_d on the field intensity in
germanium, determined experimentally by Gunn [36], is presented
in Fig. 2.2. In an electric field $E \approx 10^4$ V/cm, which is typical of
the space-charge region for a reverse voltage across a diode, the
value of v_d is $5 \cdot 10^6$ cm/sec. Scott has solved Eq. (1.13) using an
initial distribution given by Eq. (2.14) and the condition that the
velocity of holes at the boundary is finite and equal to 10^5–10^6 cm/sec.
The results of his calculations for a particular diode are presented
in Fig. 2.3. Unfortunately, Scott's paper [35] does not give analytic
expressions which have been used to plot these graphs and, therefore, general conclusions cannot be drawn from his work.

The maximum reverse current can easily be shown to be $i_{max} \approx qp_iSv_d$. Using Eq. (1.25), we obtain

$$\frac{i_{max}}{i_f} \simeq \frac{L_p v_d}{D_p} \qquad (2.10)$$

and when $L_p = 0.02$ cm, $D_p = 44$ cm^2/sec, $v_d = 10^5$–10^6 cm/sec, we find that $i_{max}/i_f = 50$–500.

Thus, allowance for the finite velocity of holes also shows that the reverse current peak is limited but it is very large (tens or hundreds of times larger than the forward current).

In practice, the value of the transient reverse current is limited by a load resistance R_l, which is always finite.*

4.3. Nonstationary Forward Current

In those cases when we cannot assume that the forward current through a diode has been flowing for an infinitely long time before switching (i.e., the forward current can be regarded as a "short"), the initial condition in the solution of Eq. (1.13) is the distribution given by Eqs. (1.55) or (1.63).

As before, we shall assume that when a forward bias is applied, a diode is connected to a current generator circuit, i.e., a constant forward current i_f flows through the circuit. The general solution of the problem, with the initial condition $p(X, \mathcal{J}_f)$, given by Eq. (1.55) and with the boundary condition (1.22), yields the following expression when the source function for a semi-infinite region [19] is used and the value of p_{n0} is neglected:

$$\frac{\partial p}{\partial X}\bigg|_{X=0} = \frac{e^{-\mathcal{J}}}{2\sqrt{\pi \mathcal{J}}} \cdot \int_0^\infty p_1 \psi_1(X, \mathcal{J}_f) e^{-\frac{X^2}{4\mathcal{J}}} X \, dX. \qquad (2.11)$$

We shall use Eq. (1.52) to make the substitution:

$$\psi_1(X, \mathcal{J}_f) = e^{-X} - \psi_4(X, \mathcal{J}_f).$$

---

* The influence of the resistance R_l on the transient switching process will be considered in the next section (§ 5).

The integral (2.11) for $p(X, \mathcal{J}_f) = \psi_4(X, \mathcal{J}_f)$ has been considered in detail in [30], where it is shown that for high values of $\mathcal{J}$ we have

$$\int\limits_0^\infty \psi_4(X, \mathcal{J}_f)\, e^{-\frac{X^2}{4\mathcal{J}}} X\, dX = p_1 \frac{e^{-\mathcal{J}}}{2\sqrt{\pi \mathcal{J}^3}} \left[\frac{2\mathcal{J}_f^{1/2} e^{-\mathcal{J}_f}}{\sqrt{\pi}} + \text{erfc}\sqrt{\mathcal{J}_f}\; \right]. \quad (2.12)$$

We note that when the same condition applies, we obtain

$$\int\limits_0^\infty p_1 e^{-X} e^{-\frac{X^2}{4\mathcal{J}}} X\, dX = p_1 \frac{e^{-\mathcal{J}}}{2\sqrt{\pi \mathcal{J}^3}}, \quad (2.13)$$

and, using Eq. (1.29), we can obtain the following expression for the reverse current density [37]:

$$j(t) \simeq j_f\; \frac{e^{-\mathcal{J}}}{2\sqrt{\pi \mathcal{J}^3}} \left[1 - \left(\frac{2\sqrt{\mathcal{J}_f}\; e^{-\mathcal{J}_f}}{\sqrt{\pi}} + \text{erfc}\sqrt{\mathcal{J}_f}\; \right) \right]. \quad (2.14)$$

Numerical estimates show that when $\mathcal{J} \geqslant 3$, the error in the determination j(t) by means of Eq. (2.4) does not exceed 20% compared with the exact solution.

In the expression (2.14), the term in front of the square brackets is in fact Eq. (2.6) and it describes the decay of the reverse current when $\mathcal{J}_f \to \infty$; the expression in the square brackets (we shall denote it by K_1) is a correction for the finite duration of the flow of the forward current.

Thus, the nature of the time dependence of the transient reverse current at high values of $\mathcal{J}$ is independent of the duration of the forward current flowing before the switching takes place, with the exception of a change in the value of the coefficient K_1. Figure 2.4 shows the dependence of K_1 on $\mathcal{J}_f$. It is evident from this figure that the distribution of holes during the flow of the forward current can be assumed to be stable after $\mathcal{J}_f > 2 - 3$. Naturally, $K_1 \to 1$ when $\mathcal{J}_f \to \infty$. To reduce the transient reverse current to a negligibly small value at some moment $\mathcal{J}$ [as is usual in radio engineering, we shall assume this value to be 0.1j ($\mathcal{J}, \mathcal{J}_f \to \infty$), the duration of the forward current pulse must be shorter than $0.3\tau_p$.

An asymptotic expression for K_1 for very short durations of the forward current pulse is

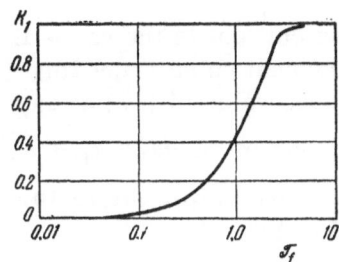

Fig. 2.4. Dependence of the coefficient K_1 on the duration of the forward current pulse

$$K_1 \simeq 0.75 \mathcal{J}_f \sqrt{\mathcal{J}_f}, \qquad \mathcal{J}_f \ll 1. \tag{2.15}$$

Our solution can be represented in the relatively simple analytic form of Eq. (2.14) only after a certain time interval has elaped since switching ($\mathcal{J} > 3$). An expression for $j(\mathcal{J}, \mathcal{J}_f)$ has been obtained in [38] for any values of $\mathcal{J}$ but this expression is cumbersome and includes integrals which can only be solved numerically.

Therefore, it is more convenient to analyze the initial stage of the transient process by means of the approximate distribution (1.63) which is valid for sufficiently short forward current pulses ($\mathcal{J}_f < 0.5$).

Without completely solving the diffusion equation (1.13), we shall use the operator method [21] to find the value of the hole concentration gradient at the p–n junction:

$$\frac{\partial p}{\partial X}\bigg|_{X=0} = \frac{2p_1 \sqrt{\mathcal{J}_f}}{\cdot \sqrt{\pi}} \left\{ \frac{e^{-\mathcal{J}}}{\sqrt{\pi \mathcal{J}}} - \frac{2}{\sqrt{\pi \mathcal{J}_f}} \times \right.$$
$$\left. \times \exp\left[-\mathcal{J}\left(1 - \frac{4}{\pi \mathcal{J}_f}\right)\right] \operatorname{erfc} \frac{2\sqrt{\mathcal{J}}}{\sqrt{\pi \mathcal{J}_f}} \right\}. \tag{2.16}$$

In general, Eq. (2.16) cannot be separated, like (2.14), into two factors, one of which is only a function of $\mathcal{J}_f$ and the other is only a function of $\mathcal{J}$. Only at very low values of $\mathcal{J}$ ($\mathcal{J} \ll \mathcal{J}_f$) can we obtain the following expression by expanding in series the exponential function and the error function and using Eqs. (1.25) and (1.29):

$$j(\mathcal{J}, \mathcal{J}_f) = j_f \frac{2}{\sqrt{\pi}} \sqrt{\mathcal{J}_f} \cdot \frac{1}{\sqrt{\pi \mathcal{J}}}. \tag{2.17}$$

We see that the decay of the reverse current is the same as in the case of an infinitely long forward current pulse preceding switching [cf. Eq. (2.5)] but the instantaneous value of $j(\mathcal{J})$ at any given moment is $\dfrac{\sqrt{\pi}}{2} \dfrac{1}{\sqrt{\mathcal{J}_f}}$ times smaller than in the case $\mathcal{J}_f \to \infty$. . For $\mathcal{J} = 0.1\mathcal{J}_f$, the expression (2.17) is satisfied only to within 40% and for $\mathcal{J} = 0.01\mathcal{J}_f$, it is satisfied to within 17%.

To determine the recovered charge in the case of a nonstationary distribution of holes during the flow of the forward current, we shall integrate Eq. (2.16) with respect to time between the limits 0 and ∞.

Integration on the assumption that $\mathcal{J}_f \ll 1$ gives

$$Q_{\text{rec}} = S \int_0^\infty j(\mathcal{J}, \mathcal{J}_f)\, dt \simeq i_f\, t_f \left(1 - \frac{\sqrt{\pi}}{2} \sqrt{\mathcal{J}_f}\right). \qquad (2.18)$$

The product $i_f t_f$ is simply the charge transported by the forward current pulse. Thus, when the duration of the forward current pulse is reduced, the value of the charge flowing out of the base after the switching approaches the value of the injected charge. Physically, this conclusion is self-evident because the recombination of holes in the base can be neglected when $\mathcal{J}_f \ll 1$.

However, it is interesting that the charge lost by recombination is proportional not to $\mathcal{J}_f$, as is often assumed, but to $\sqrt{\mathcal{J}_f}$. For 90% of the injected charge to flow back (after the switching) into the external circuit, the condition $t_f \lesssim 0.01\tau_p$ must be satisfied rather than the condition $t_f < 0.1\tau_p$.

On the other hand, when $\mathcal{J}_f \ll 1$, we can restrict ourselves to terms of the first order of smallness in Eq. (2.18) and then we obtain

$$Q_{\text{rec}}(\mathcal{J}_f) \simeq Q_{\text{st}} \cdot \mathcal{J}_f \qquad (2.19)$$

or, using Eq. (2.8), we obtain

$$Q_{\text{rec}}(\mathcal{J}_f) \simeq Q \ (\mathcal{J}_f \to \infty) \cdot 2\mathcal{J}_f \qquad (2.20)$$

Thus, a reduction of the forward current pulse duration has a different effect on the various phases of the switching:

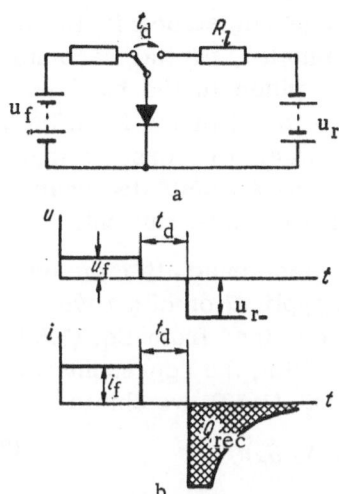

Fig. 2.5. Transient processes during the switching of a semiconductor diode from the forward to the reverse direction with a delay time between the forward and reverse pulses: a) switching circuit; b) time dependences of the voltage and the current in a circuit containing a diode and a load resistance R_l.

1) the instantaneous values of $j(\mathcal{J}, \mathcal{J}_f)$ during the final phase of the transient process $(\mathcal{J} > 2 - 3)$ decrease proportionally to the term $0.75 \mathcal{J}_f \sqrt{\mathcal{J}_f}$;

2) the instantaneous values of $j(\mathcal{J}, \mathcal{J}_f)$ during the initial phase of the transient process $(\mathcal{J} \ll 0.01 \mathcal{J}_f)$ decrease proportionally to $1.13 \sqrt{\mathcal{J}_f}$;

3) the recovered charge, which is an integral characteristic of the transient process, decreases by about $2 \mathcal{J}_f$ compared with the values of the charge for an infinitely long forward current pulse.

4.4. Delayed Switching

Sometimes there is a finite delay between the end of the forward current pulse and the beginning of the reverse voltage pulse. This problem has been considered by Nosov [39].

The diode switching circuit and the time dependences of the voltage and the current are given in Fig. 2.5. In analyzing this case, we shall assume, as before, that $R_l = 0$ and that a forward current of density j_f has been flowing through the diode for an infinitely long time before the switching.

The flow of the forward current through the base results in the accumulation of excess holes; during the delay stage (t_d) neither the forward nor the reverse current flows through the diode (this corresponds to an infinitely large resistance in the diode circuit in the absence of external signals), partial recombination of the accumulated charge takes place in the base and the densities of holes become equalized by the diffusion from the p-n junction into the interior of the base; when a reverse voltage pulse is applied, the process of dispersal of the accumulated charge begins due to the flow of a large transient reverse current.

The distribution of holes in the diode base after the end of the forward current pulse and before the application of a reverse voltage pulse (during the delay stage) is obtained from Eq. (1.50) by substituting into that equation $j_0 = 0$. Using the representation given by Eq. (1.54) and bearing in mind Eq. (1.52), we obtain

$$p(X, \mathcal{J}_d) = p_1 \psi_4(X, \mathcal{J}_d), \qquad (2.21)$$

where $\mathcal{J}_d = t_d/\tau_p$ is the dimensionless delay time, measured from the moment when the forward current ceases to flow.

The expression (2.21) is the initial condition for the solution of the diffusion equation after the application of the reverse voltage. The boundary condition is, as before, Eq. (1.22). The expression for the reverse current, obtained using the source function for a semi-infinite line, is of the form

$$j(\mathcal{J}_d, \mathcal{J}) = qD_pL_p\left(\frac{\partial p}{\partial X}\right)_{X=0} = j_f \frac{e^{-\mathcal{J}}}{2\sqrt{\pi\mathcal{J}^3}} \int_0^\infty \psi_4(X, \mathcal{J}_d)e^{-\frac{X^2}{4\mathcal{J}}} X\,dX, \qquad (2.22)$$

where $\mathcal{J}$ is measured from the end of the delay stage. The general form of the integral in Eq. (2.22) is not used because at low and high values of $\mathcal{J}$ we can employ the following asymptotic approximations [30]:

$$j(\mathcal{J}_d, \mathcal{J}) \simeq j_f \text{ erfc }\sqrt{\mathcal{J}_d}\left(\frac{e^{-\mathcal{J}}}{\sqrt{\pi\mathcal{J}}} + \text{erf}\sqrt{\mathcal{J}}\right), \quad \mathcal{J} \ll 1, \qquad (2.23)$$

$$j(\mathcal{J}_d, \mathcal{J}) \simeq j_f \frac{e^{-\mathcal{J}}}{2\sqrt{\pi\mathcal{J}^3}}(1 - K_1), \quad \mathcal{J} \gg 1, \qquad (2.24)$$

where K_1 is the same coefficient as in Eq. (2.14), but we must remember to replace $\mathcal{J}_f$ with $\mathcal{J}_d$. Comparison of Eqs. (2.5) and (2.6)

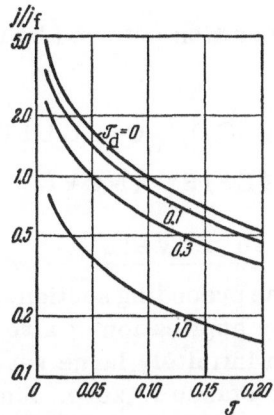
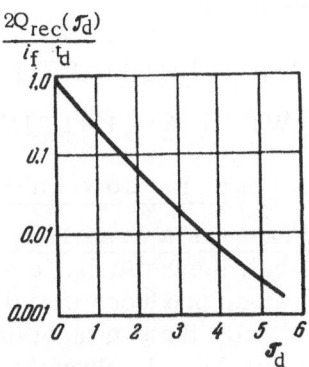

Fig. 2.6. Decay of the tran-   Fig. 2.7. Dependence of the
sient reverse current for various recovered charge on the de-
durations of the delay time.   lay time.

shows that the nature of the decay of the transient reverse current
is the same as in the absence of any delay stage; the curve $j(\mathcal{J})$ is
simply altered by a factor of $\dfrac{1}{\text{erfc}\,\sqrt{\mathcal{J}_d}}$ in the initial phase and by a
factor $1/(1 - K_1)$ in the final phase of the transient process. Figure
2.6 shows the $j(\mathcal{J})$ curves plotted using the exact formula (2.22) for
several values of the delay time.

To calculate the recovered charge, using the integral in Eq.
(2.22), we shall replace, in accordance with Eq. (1.52), the function
$\psi_4(X,\ \mathcal{J}_d)$ by $e^{-x} - \psi_1(X,\ \mathcal{J}_d)$, and we shall use an approximate substitution for $\psi_1(X, \mathcal{J}_d)$ using the left-hand side of Eq. (1.63). Car-
rying out the successive integration with respect to the variables
X and $\mathcal{J}$ between the limits 0 and ∞, we obtain [39]:

$$Q_{\text{rec}}(\mathcal{J}_d) \simeq \frac{i_f\,\tau_p}{2}\,\text{erfc}\,\sqrt{\mathcal{J}_d}\left(1 - \frac{\text{erfc}\,\sqrt{\mathcal{J}_d}}{2e^{-\mathcal{J}_d}}\right). \qquad (2.25)$$

This function is shown in Fig. 2.7. The curve in Fig. 2.7 can be
approximated by the following exponential function

$$2Q_{\text{rec}}(\mathcal{J}_d) \simeq \frac{Q_{st}}{2}\,e^{-1.2\mathcal{J}_d},$$

which, in the most interesting range of values of $2Q_{\text{rec}}(\mathcal{J}_d)/i_f\tau_p$
between 0.5 and 0.005, is practically identical with the solution

given by Eq. (2.25) and with the exact solution obtained by numerical integration [41].

§5. SWITCHING OF A DIODE CIRCUIT
WITH A LIMITING RESISTANCE

5.1. Low Injection Levels

Two Phases of a Transient. In the preceding section, we have pointed out that, theoretically, a sudden application of a reverse voltage to a diode produces initially an infinitely large reverse current, i.e., initially the p-n junction resistance is zero. Since in any real circuit there is always a load resistance in series with a diode, the maximum value of the transient reverse current is limited by this resistance.* Thus, for a certain times after the switching the value of the reverse current is governed only by the parameters of the external circuit and is independent of the properties of the diode itself. This situation continues until the distribution of holes in the base changes so that the resistance of the p-n junction is no longer equal to zero. During this period, the reverse current through the diode is practically constant and equal to

$$i_0 \simeq \frac{U_r}{R_l} .$$
                                                          (2.27)

The voltage across the p-n junction during this period decreases from the value u_f, corresponding to the forward current i_f, to zero. Therefore, Eq. (2.27) is reasonably accurate when the value of R_l is large compared with the base resistance R_b† and when the amplitude of the reverse pulse U_r is large compared with the value of u_f.

The constancy of the reverse current means that in the integration of the diffusion equation we must use the boundary condition of the second kind given by Eq. (1.20). The solution of this problem, given by Eq. (1.50), has been obtained in § 3; the distributions of the hole density in the base at various times are shown

---

* The value of the load resistance includes, in addition to the load resistance proper, two other components: the resistance of the interior of the crystal diode and its ohmic contacts, as well as the output resistance of the generator of the reverse voltage pulses. Each of these components cannot be reduced below serveral ohms.

† The value of R_b should be small to make i_0 constant because the value of R_b itself changes during a transient process.

graphically in Fig. 1.2. From these distributions, it follows that
the boundary condition (1.20) is satisfied until the hole density at
the p–n junction decreases to zero. In order to maintain a constant
value of $(\partial p/\partial X)_{X\,=\,0}$ beyond this stage, it is necessary further to
reduce p(0, t) so that it acquires negative values, which is physi-
cally impossible.

Therefore, when the density of holes at the p–n junction decreases to zero, the density gradient begins to decrease and this
is equivalent to a decrease of the reverse current, which then tends
asymptotically to the saturation current i_s.

Thus, the whole switching transient of a diode in a circuit
with a limiting resistance can be divided into two phases: the initial phase during which the boundary condition (1.20) applies and
the reverse current (governed solely by the circuit elements) remains quasiconstant; this is followed by the final phase during
which the boundary condition (1.22) applies and the reverse current
decreases in accordance with a law governed by the characteristics
of the diode. Theoretically, the second phase is infinitely long but
in practice we may assume that this phase is complete when the
reverse current falls to some specified value i_{rec} (known as the
recovery current). Therefore, when we speak of the duration of the
second phase, we assume that it is finite.

In the American and British literature the first and second
phases are usually called the recovery and reverse phases: we
shall use these terms here; more recently, the initial and final
stages of a reverse switching transient have been called the storage
and transition phases, respectively.

In the Soviet literature the first phase is called the high-re-
verse-conductance phase (faza vysokoi obratimoi provodimosti) or,
as suggested in Nosov's review [40], the "plateau" (polochka) because of the plateau-like appearance of the relevant time dependence of the current. Also in the Soviet literature, the second phase
is called the reverse-current (or reverse-resistance) recovery
phase [faza vosstanovleniya obratnogo toka (soprotivleniya)]. Frequently, the whole transient process is referred to as the reverse-
current recovery phase (particularly in engineering literature);
this definition suppresses the distinction between the natures of the
flow of the reverse current during the two phases.

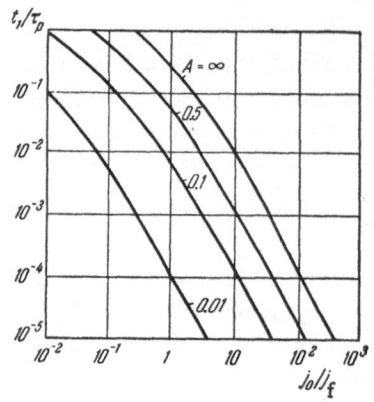

Fig. 2.8. Dependences of the recovery phase duration on the switching conditions for various radii ($A = r_0 | L_p$) of a p-n junction.

Duration of the Recovery Phase. The duration of the first (recovery) phase t_1 can be easily found from Eq. (1.50) by assuming that $p(X, \mathcal{J}) = 0$ for $X = 0$ and $\mathcal{J} = \mathcal{J}_1$ ($\mathcal{J}_1 = t_1/\tau_p$). After performing fairly simple transformations, we obtain

$$\operatorname{erf} \sqrt{\mathcal{J}_1} = \frac{j_f}{j_f + j_0} = \frac{1}{1 + B}, \tag{2.28}$$

where $B = j_0/j_f = i_0/i_f$. This classically simple expression, relating t_1/τ_p with j_0/j_f was deduced simultaneously by Kingston [8] and by Lax and Neustadter [9].

The dependence of the duration of the recovery phase on the swithing conditions can be seen from Fig. 2.8 (a planar diode is represented by the curve with $A = \infty$). It is interesting to note that the duration of the recovery phase t_1 is proportional to the hole lifetime τ_p and that it depends on the switching conditions only through the ratio of the current densities (or currents) B.

It is interesting to note that throughout the recovery phase the voltage across the p-n junction remains positive. In fact, following Eq. (1.21), we obtain

$$u_{p-n}(\mathcal{J}) = \frac{kT}{q} \ln \frac{p(0, \mathcal{J})}{p_{n0}}. \tag{2.29}$$

During the time when $p(0, \mathcal{J})$ decreases from p_1 to p_{n0}, the value of $u_{p-n}(\mathcal{J})$ decreases from u_f (corresponding to the current i_f) to zero but it remains positive; when $p(0, \mathcal{J})$ decreases from p_{n0} to

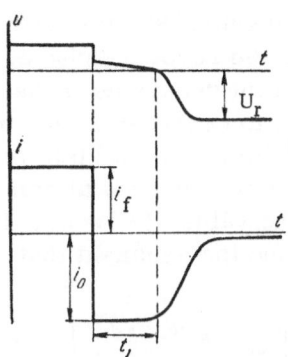

Fig. 2.9. Time dependences of
the voltage across a diode and
the current through it during a
transient process for $R_l \neq 0$.

zero, the voltage across the p-n junction changes its polarity and increases to U_r. The time dependences of the current through the diode and of the voltage across it during the whole transient process are shown in Fig. 2.9.

Reverse Phase. To calculate the time dependence of the reverse current after the end of the recovery phase (i.e., when $\mathcal{J} > \mathcal{J}_1$), it is necessary to solve the diffusion equation with new boundary and initial conditions: namely

$$p(0, \mathcal{J}) = 0 \quad \text{when} \quad \mathcal{J} \geqslant \mathcal{J}_1, \qquad (2.30)$$

$$p(X, \mathcal{J}_1) = p_1 \left[e^{-X} - \frac{1}{\text{erf} \sqrt{\mathcal{J}_1}} \psi_1(X, \mathcal{J}_1) \right]. \qquad (2.31)$$

The condition (2.30) is identical with Eq. (1.22) applied at a different time and the distribution (2.31) can be easily obtained from Eq. (1.50) by substituting into the latter $\mathcal{J} = \mathcal{J}_1$ and using Eq. (2.28).

The time dependence of the reverse current under these conditions is calculated in the same manner as in the analysis of Eq. (2.11) because $\psi_1(X, \mathcal{J})$ and $\psi_4(X, \mathcal{J})$ are related by a simple expression given by Eq. (1.52). As before, an exact analytic expression cannot be obtained and all we can do is numerically to integrate the general solution (1.13). The time dependences of the reverse current (the time $\mathcal{J}$ is measured from the end of the recovery phase) calculated for two values of $\mathcal{J}_1$, are given in Fig. 2.10.

However, there is another way of obtaining an approximate equation describing the decay of the reverse current, which was

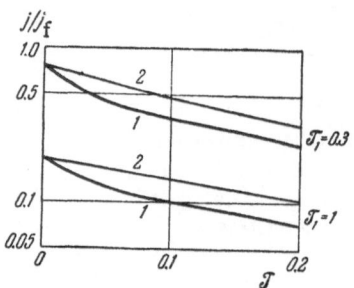

Fig. 2.10. Decay of the reverse current after the end of the recovery phase, calculated using exact (1) and approximate (2) formulas for two different values of $\mathcal{J}_1$.

used by Kingston [8]. He assumed that, after the recovery phase, the reverse current decays as if that current began to decrease from an infinite value (i.e., from a state in which $R_l = 0$) and has reached at some moment $\mathcal{J}'$ the value $i(\mathcal{J}') = i_0$. In other words, Kingston assumed that

$$\left.\frac{\partial p(X, \mathcal{J}_1)}{\partial X}\right|_{\substack{X=0, \\ R_l \neq 0}} = \left.\frac{\partial p(X, \mathcal{J}')}{\partial X}\right|_{\substack{X=0 \\ R_l=0}}, \quad (2.32)$$

and he postulated that this equation is valid for $\mathcal{J} > \mathcal{J}_1$ for the function on the left and for $\mathcal{J} > \mathcal{J}'$ for the function on the right. Thus, a general description of the transient process consists of a simple combination of the solution (2.28) for the recovery phase and of the solution (2.4), which is approximately valid for the reverse phase.

To check the validity of this assumption, Fig. 2.11 shows the distributions of the hole densities in the base calculated for $\mathcal{J} = \mathcal{J}_1$ using Eq. (2.28) and for $\mathcal{J} = \mathcal{J}'$ using Eq. (2.4) for two different values of $\mathcal{J}_1$. Examination of the curves in Fig. 2.11 shows that the approximate and true distributions are indeed close and that for any value of X the quantity $p(X, \mathcal{J}')$, obtained from Eq. (2.4) is larger

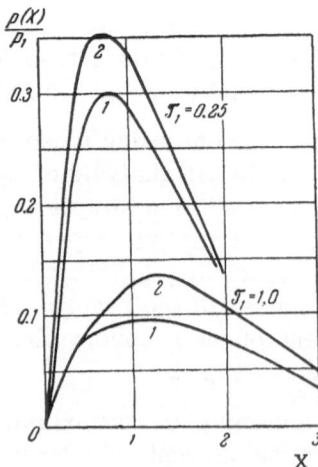

Fig. 2.11. Distribution of the holes in the base at the end of the recovery phase, calculated using exact (1) and approximate (2) formulas for two different values of $\mathcal{J}_1$.

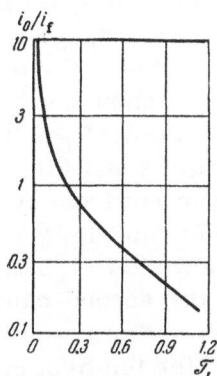

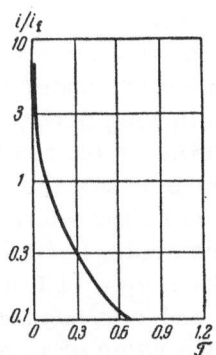

Fig. 2.12. Dependence
of the recovery phase
duration on the switch-
ing conditions.

Fig. 2.13. Dependence
of the transient reverse
current on time.

than $p(X, \mathcal{J}_1)$, calculated from Eq. (2.28) for the same value of B.
Naturally, the larger value of B and, consequently, the shorter the
duration of the recovery phase $\mathcal{J}_1$, the smaller is the relative difference between the approximate and true hole distributions. Thus,
we may expect the transient reverse current calculated on the basis of Kingston's assumption to be somewhat higher than the true
value but the difference between the calculated and true values
should decrease with decreasing effect of the load resistance R_l
during the recovery phase.

Figure 2.10 shows the approximate Kingston's solutions for
the same values of $\mathcal{J}_1$, for which exact dependences have been calculated by Henderson and Tillman [30]. We can see that the nature
of the decay of the reverse current is the same in both cases (with
the exception of a short interval immediately after the end of the
recovery phase) and only the instantaneous values of $j(\mathcal{J})$ are approximately 30% higher in the approximate solution than their true
values.

To calculate rapidly the duration of a transiwnt under various
switching conditions, it is convenient to use the curves given in Figs.
2.12 and 2.13. For the sake of illustration, we shall consider a specific example. Let us assume that, when a certain planar diode is
switched from a forward current i_f = 10 mA by a reverse step voltage

U_r = 10 V in a circuit containing a load resistance R_l = 1 k Ω, the time taken for the reverse current to decrease to i_{rec} = 5 mA is 4 μsec. The question is what will be the duration of the transient process in the same diode under different switching conditions represented by the following parameters: i_f = 5 mA; U_r = 3 V; R_l = 1 k Ω; i_{rec} = 0.5 mA. In the first case i_0/i_f = 1 and Fig. 2.12 shows that $\mathcal{J}_1$ = 0.23. The decay of the reverse current from 10 to 5 mA takes a time $\mathcal{J}_2$ = 0.20—0.09 = 0.11 , as found from Fig. 2.13. According to the measurements, in the first case we have t_1 + t_2 = 4 μsec = 0.34τ_p and, therefore, τ_p = 11.0 μsec. In the second case, i_0/i_f = 0.6, i_{rec}/i_f = 0.1 and correspondingly $\mathcal{J}_1$ = 0.39 , $\mathcal{J}_2$ = 0.66—0.16 = 0.50. Using the value of τ_p calculated for the first case, we find that t_1 + t_2 = 0.89τ_p = 10.4 μsec.

The exact effect of the load resistance on the nature of the transient process can be found only by numerical integration of the general solution of Eq. (1.13). The results of such a calculation for a specific germanium diode are given in [9]; it is assumed that the parameters of this diode are S = 2.5 $\cdot$ 10^{-3} cm^2; p_{n0} = 2.5 $\cdot$ 10^{12} cm^{-3}; τ_p = 100 μsec; i_f = 6 mA; U_r = 6 V.

The graphs describing the transient processes in this diode (Fig. 2.14) show that the total time for the decay of the reverse current to the low values represented by i_{rec}, measured from the

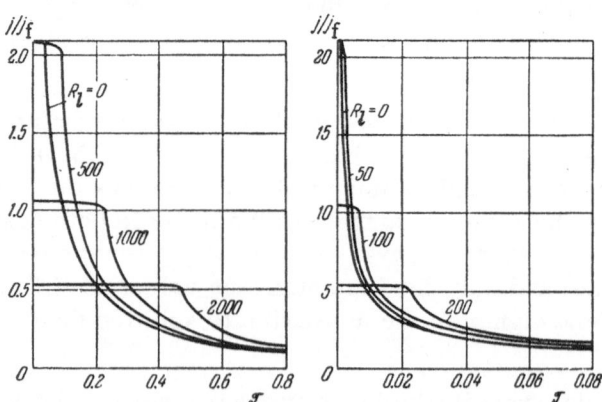

Fig. 2.14. Dependence of the transient reverse current on time for various load resistances R_l (in ohms). The scales in the two parts of the figure are different.

beginning of the transition, is almost independent of the value of R_l if $i_{rec} \leq (0.1 - 0.2)i_0$.

Role of Barrier Capacitance. We shall now estimate the influence of the p-n junction barrier capacitance on the transient processes. In the equivalent circuit of the diode, this capacitance is assumed to be connected in parallel with the p-n junction so that the current through it is given by the expression

$$i_C = C_d \frac{du_{p-n}}{dt} + u_{p-n} \frac{dC_d}{dt}. \tag{2.33}$$

The highest capacitative current flows at the end of the recovery phase since at this time the derivative du_{p-n}/dt has its maximum value (cf. Fig. 2.9). If we assume, in the first approximation, that the diode capacitance is constant, we find, using the expression (2.48) for u_{p-n} obtained in § 6,

$$\frac{i_C}{i_0} = \frac{R_l C_d}{U_r \tau_p} \cdot \frac{kT}{q} \cdot \frac{i_f + i_0}{i_s} \cdot \frac{e^{-\mathcal{J}_1}}{\sqrt{\pi \mathcal{J}_1}}, \tag{2.34}$$

where i_s is the saturation current given by

$$i_s = S \frac{p_{n0} D_p q}{L_p}. \tag{2.35}$$

It is evident from Eq. (2.34) that the role of the barrier capacitance becomes less important when the value of the hole lifetime τ_p increases and the value of C_d decreases. The dependence of i_C/i_0 on R_l is given explicitly in Eq. (2.34) because, other conditions being constant, the value of $\mathcal{J}_1$ is a function of R_l. Using Eq. (2.28), we can easily show that when R_l is increased, the factor $\frac{e^{-\mathcal{J}_1}}{\sqrt{\pi \mathcal{J}_1}}$ decreases much more rapidly so that the role of the barrier capacitance decreases.

Short Forward Current Pulses. The solution of the diffusion equation with the boundary condition (1.20) for a nonstationary distribution of holes in the case of a forward current, which is described by Eq. (1.55) under the current generator conditions, was obtained by Ko [27]. The duration of the recovery phase is given by the formula

$$\text{erf} \sqrt{\mathscr{J}_1} = \frac{1}{1+B} \text{erf} \sqrt{\mathscr{J}_f + \mathscr{J}_1},$$

(2.36)

which reduces to Eq. (2.28) when $\mathscr{J}_f > 1$. When $\mathscr{J}_f \ll 1$, it follows automatically that $\mathscr{J}_1 \ll 1$; then, using the approximation for the error function in the case of small values of the argument, given by Eq. (1.57), we obtain

$$\mathscr{J}_1 = \frac{1}{B} \frac{\mathscr{J}_f}{2+B}, \quad \mathscr{J}_f \ll 1,$$

(2.37)

and when we assume an additional condition $B \ll 1$,

$$t_1 = \frac{1}{2} t_f \frac{i_f}{i_0}.$$

(2.38)

It is interesting to note that this relationship was predicted by Gossick in 1953 [10] without any theoretical calculations or experimental studies. On the other hand, when $B \gg 1$, it follows from Eq. (2.37) that

$$t_1 \simeq \left(\frac{i_f}{i_0}\right)^2 t_f.$$

(2.39)

Calculation of the decay of the reverse current after the recovery phase has not been carried out for a short forward current pulse but physical considerations show that this case should not differ greatly from the case of a steady-state forward current (before switching).

Switching After Delay. Let us assume that switching takes place in the circuit shown in Fig. 2.5 and that $R_l \neq 0$. The initial distribution of the hole density in the base at the moment of arrival of a reverse voltage pulse is given by Eq. (2.21). The solution of the diffusion equation, with the boundary condition given by Eq. (1.20), yields [39]

$$\text{erf} \sqrt{\mathscr{J}_1} = \frac{j_f}{j_0} \text{erfc} \sqrt{\mathscr{J}_1 + \mathscr{J}_d},$$

(2.40)

where the time $\mathcal{J}_1$ is measured from the end of the delay. The dependence of the duration of the recovery phase on the delay time, calculated using Eq. (2.40) for various values of the parameter $B = j_0/j_f$, is shown in Fig. 2.15. When $\mathcal{J}_d \to 0$, Eq. (2.40) reduces to Eq. (2.28), but when $\mathcal{J}_d \gg 1$ and $\mathcal{J}_1 \ll 1$ an expansion of the error function into a series gives

$$\mathcal{J}_1 \simeq \frac{1}{4}\left(\frac{i_f}{i_0}\right)^2 \frac{e^{-2\mathcal{J}_d}}{\mathcal{J}_d}, \qquad (2.41)$$

i.e., for large values of the delay time the duration of the recovery phase decreases almost exponentially with increasing $\mathcal{J}_d$.

5.2. High Injection Levels

In Chap. I, we mentioned that at sufficiently high injection levels $\left(\Delta = \dfrac{p(x,t)}{n_{n0}} \gg 1\right)$ the system of equations (1.3) – (1.8) can be made linear in the same way as for low injection levels.

However, analysis of transient processes in a planar diode with a semi-infinite base does not yield a linear differential equation for the switching of large currents. This is because when the condition $\Delta > 1$ applies to the flow of the forward current through the p-n junction, the base has a whole spectrum of injection levels (from maximum to zero value) due to the decrease of the density of the excess holes away from the junction. Therefore, we can solve numerically the complete nonlinear equation (1.15) only for some special cases. In these cases, the condition $\Delta \gg 1$ is not essential and the switching problem can be solved for an arbitrary injection level.

It is particularly important to obtain a solution which allows for the influence of high injection levels during the recovery phase

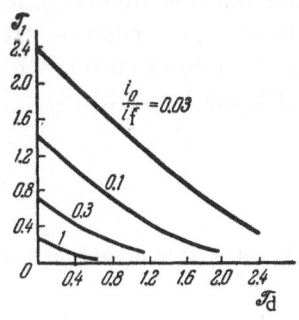

Fig. 2.15. Dependence of the recovery phase duration on the delay time for various values of the parameter i_0/i_f.

of the transient process. This is because the formula (2.28) is used extremely widely in the determination of the carrier lifetimes in semiconductors, particularly in studies of the dependence of τ_p on the injection level. Therefore, the derivation of a dependence similar to Eq. (2.28) for arbitrary values of the injection level is of great practical importance.

Analysis of the second (reverse) phase of the transient process at high injection levels is not very important because in the majority of the switching conditions the density of holes at the end of the first (recovery) phase is much lower than at the beginning of this phase so that the condition $\Delta < 1$ is satisfied over a large part of the base. This is why there have been no special investigations of the reverse phase of the transient processes at high injection levels.

Duration of the Recovery Phase. The fullest analysis of the first (recovery) phase of the transient process has been carried out by Iglitsyn, Kontsevoi, et al. [42-45]. They have analyzed the case of moderate injection levels $(0.1 < \Delta < 10)$* retaining the assumptions made in Chap. I (one-dimensional problem for a semi-infinite region in which the electrical neutrality condition is obeyed; the injection efficiency is assumed to by $\gamma = 1$).

In the physical sense, the main features of such switching conditions are affected by two factors. Firstly, when $p(0) \gtrsim n_{n_0}$, a driving (i.e., accelerating the motion of holes away from the p-n junction) electric field appears in the base [46]. Secondly, when the condition $\Delta \ll 1$ is not satisfied, the Shockley-Read theory predicts that the lifetime is no longer constant. These two factors result in the driving of holes away from the junction.

The influence of the driving field is allowed for by the structure of Eq. (1.15), as indicated by the presence of terms containing $\partial p/\partial x$. It is also assumed that the lifetime is governed by steady-state recombination at simple singly charged centers (cf. Chap. VII), i.e., that it varies in accordance with the law [13]:

$$\tau_p = \tau_0 \frac{1 + a\Delta}{1 + c\Delta},$$
(2.42)

---

* The numerical value of the injection level is the value of Δ at the space-charge region boundary near the p-n junction.

where $a/c = \tau_\infty/\tau_0$ and τ_∞, τ_0 are, respectively, the hole lifetimes for infinitely high and vanishingly low injection levels.

The solution of the problem is given in [43, 44] for the following special cases:

$$j_0/j_f = 1, \qquad \tau_\infty/\tau_0 = 0.1 - 10, \qquad \Delta = 0.1 - 10.$$

A numerical calculation, carried out using an electronic computer, has shown that when the lifetime is constant (i.e., when $\tau_\infty/\tau_0 = 1$), the presence of a driving field pushes holes into the interior of the base but, as in the case of low injection levels, the distribution of holes during the flow of a constant forward current obeys an exponential law of the type

$$p(x) = p(0) \exp\left[-\frac{x}{\alpha(\Delta) L_p}\right], * \qquad (2.43)$$

where $\alpha(\Delta)$ is some coefficient which depends on the injection level and varies from 1 (for $\Delta \ll 1$) to $\alpha = \sqrt{2b/(\alpha+1)}$ (for $\Delta \gg 1$). This result is in agreement with the general conclusion obtained by Stafeev [16] for $\Delta \gg 1$ (cf. § 2). For an arbitrary injection level and b = 2, a numerical calculation yields

$$\alpha(\Delta) = \frac{1.41}{3\Delta + 2} \sqrt{\frac{6\Delta^3 + 16\Delta^2 + 11\Delta + 2}{1 + \Delta}}. \qquad (2.44)$$

The nature of the hole density distribution in the base at different times after switching to the reverse voltage is shown in Fig. 2.16. It is important to note that throughout most of the first (recovery) phase, the hole density near the p-n junction remains practically constant and then it decreases rapidly to zero. It follows that the recovery phase duration should be governed by the hole lifetime corresponding to the injection level $\Delta = p(0)/n_{n0}$.

Analysis of the results of calculations for a wide range of values of Δ and τ_∞/τ_0 shows that the formula (2.28) can still be used in the determination of t_1 if τ_p is replaced with the lifetime at the injection level $\Delta = p(0)/n_{n0}$, corresponding to the forward

---

* Here, we cannot use the parameter p_1 because the value of $p(0)$ is not proportional to j_f due to the dependence $\tau_p = \tau_p(\Delta)$.

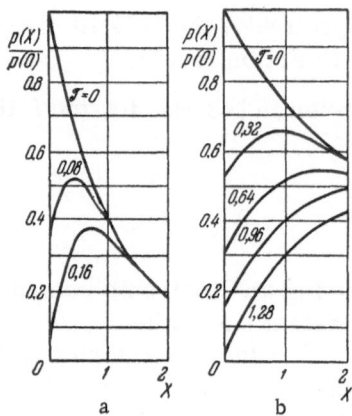

Fig. 2.16. Changes in the excess hole distribution in the base during a transient process (B = 1): a) $\tau_\infty/\tau_0 = 1$ and $\Delta = 0.5$; b) $\tau_\infty/\tau_0 = 10$ and $\Delta = 1$..

current before switching. In this case, the accuracy of the determination of $\tau_p(\Delta)$ by means of Eq. (2.28) is not less than 10%.

This result can be extended also to $j_0/j_f > 1$ because, due to the reduction of the recovery phase duration, $\mathcal{J}_1$, the change in the hole distribution near the p-n junction during the recovery phase will be even smaller than in the case $j_0/j_f = 1$.

Analysis of the relationship between the effect of the driving field on holes and the values of Δ and τ_∞/τ_0 shows that when the lifetime decreases with increasing injection level, the accumulation of excess holes takes place mainly in the contact regions. When the forward current is increased, the accumulated charge increases in magnitude and approaches the junction.

However, when the lifetime increases with increasing injection level the holes are driven into the interior of the base and the effectiveness of this phenomenon increases with increasing forward current.

Very High Injection Levels. When the injection level is increased still further, we find that two additional factors begin to influence the duration of the two phases of a switching transient and the resistance of the diode as a whole.

Firstly, when the impressed density of the holes during the flow of the forward current becomes comparable with their density in the p-type region, p_{p0} (i.e., when we can no longer assume that $p_1(0)/p_{p0} \ll 1$), we find that the injection efficiency γ decreases.

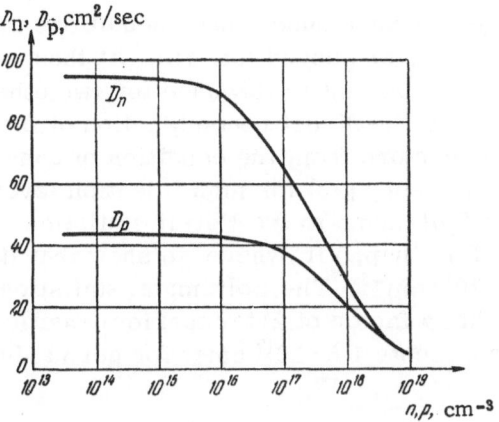

Fig. 2.17. Dependences of the hole (D_p) and
electron (D_n) diffusion coefficients in germanium on the carrier densities at T = 300°K.

In this case, we cannot ignore the injection of electrons into the p-
type region and, therefore, the boundary condition in the continuity
equation becomes nonlinear. Carriers are accumulated in the base
region as before but at high injection levels the rise of the excess
hole density with increasing forward current begins to slow down.
Consequently, we may expect a relative decrease of the duration
of the whole transient switching process.

Secondly, when the injection level is increased the carrier
mobility and the carrier diffusion coefficient decrease because of
the scattering of carriers on other carriers, so that Eq. (1.15) again
becomes nonlinear even when the third and fourth terms are dropped
from this equation. The dependences of the hole and electron diffusion coefficients on the carrier density in germanium, used by
Fletcher [47] to calculate the current–voltage characteristics of
p–n junctions at very high current densities, are shown in Fig. 2.17.
Although the curves in this figure were obtained for a simplified
spherical model of the energy bands, assuming that the constant-
energy surfaces of germanium are isotropic, these curves still represent quite satisfactorily the true dependences. The reduction
in the carrier mobility begins approximately at mobile carrier densities of the order of 10^{16} cm^{-3} and becomes so strong that it may
affect the response of a diode when p $\lesssim 10^{18}$ cm^{-3}. Such hole densities
in the base are reached only at very high current densities, which

we shall not consider because under these conditions the quasineutrality equation is no longer obeyed (cf. §2). At these very high current densities, we must make allowance for two other effects mentioned by Fletcher: the direct recombination of free carriers, which results in a departure from the condition of constancy of τ_p, predicted by the trap theory [13] for high injection levels; and the transition from the Boltzmann to the Fermi statistics. The first effect is considered in Chap. VII, where we show that it is important only when $p > 10^{18}$ cm^{-3}. The Boltzmann statistics gives correct results (to within a factor of 2) for carrier densities not greater than $4N_V$, which represents $1.5 \cdot 10^{19}$ cm^{-3} for germanium at room temperature [47].

A semiquantitative calculation of the first (recovery) phase duration was carried out by Barsukov [48, 51] making allowance for a deviation of γ from 1. Barsukov assumed that the recovery phase duration t_1 is proportional to the impressed hole density $p_1(0)$, which is obtained during the flow of a forward current of density j_f. The value of $p_1(0)$ was calculated by Barsukov for quasineutral conditions on the assumption that excess carriers in the n–type region are distributed exponentially, while in the p–type region the excess carrier density decreases linearly away from the p–n junction. Barsukov assumed that the injection level is sufficiently high to be able to assume that the hole lifetime is constant and the diffusion coefficients of holes and electrons are equal.

From general equations for the currents in the p– and n–type regions, Barsukov obtained an equation cubic in $p_1(0)$; this equation includes also the voltage drop across the p–n junction. The final formulas include such a large number of parameters, which can be varied within fairly wide limits, that Barsukov's results can be regarded only as qualitative.

The main conclusion of Barsukov's calculations is that at high forward current densities the recovery time depends not only on the ratio j_0/j_f (as in the case $\Delta \ll 1$) but also on the value of j_f: $t_1 \propto j_f^{-0.6}$. In other words, when the forward current density is increased, the response time of the diode decreases (provided $j_0/j_f = \text{const}$).

Figure 2.18 shows a calculated curve representing the relative change of the recovery time t_1 when i_f is increased (here, $i_0/i_f = 0.5$). This curve is plotted using parameters typical of

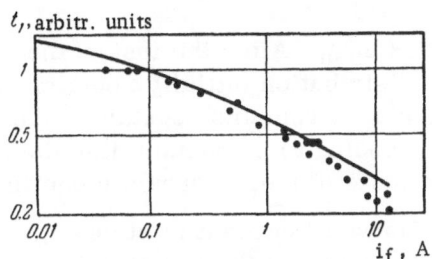

Fig. 2.18. Relative change in the recovery phase duration t_1 due to an increase of the forward current i_f. The figure shows a theoretical curve and experimental point (black dots) for DGTs21-DGTs27 diodes

alloyed germanium diodes of the DGTs$_{21}$-DGTs27 types: $n_{n0} = 2 \cdot 10^{14}$ cm^{-3}, $L_p = 2.5 \cdot 10^{-2}$ cm, $D_p = D_n = 50$ cm^2/sec, $p_{p0} = 10^{18}$ cm^{-3}, $S = 0.01$ cm^2. Although this calculation has been carried out for diodes with bases of finite thickness ($W/L_p = 1$), the slowing down of the rise of the excess current density with increasing current is not in any way related to the proximity of the ohmic contact to the p-n junction. The form of the curve in Fig. 2.18 is not affected greatly by the assumption that the base is infinitely thick ($W/L_p \gg 1$).

§ 6. SWITCHING OF A DIODE CIRCUIT WITH AN INFINITE RESISTANCE

In some cases, the conditions in a diode circuit at the end of the forward current pulse are best described by $R_l = \infty$. This case is known as the open-circuit condition.

Here, we consider the following problem. We assume that the forward current i_f has been flowing for a sufficiently long time through the diode and that the hole distribution in the base obeys Eq. (1.24). At a time $t = 0$, the diode circuit is suddenly opened so that neither the forward nor the reverse current can flow through the diode.

The excess holes can disappear from the base only by recombination. The stopping of the current in this way produces a transient relaxation of the voltage across the p-n junction. When a forward current i_f is flowing, the density of the holes near the p-n junction is higher than the equilibrium density and, consequently, according to Eq. (1.21), there is a corresponding voltage drop

across the p-n junction u_{p-n}. After the instantaneous stoppage of
the current, the hole distribution initially remains contant (and
consequently the value of u_f remains constant); the hole density
decreases with time, tending to a constant distribution $p(x) = p_{n0}$,
which represents the state of thermodynamic equilibrium.

Thus, when the forward current is stopped, we should observe
a decrease of the voltage across the p-n junction from its initial
value (observed during the flow of the forward current) to zero.
We shall now find the function $u_{p-n}(t)$ which describes the decay of
the postinjection open-circuit voltage (or the postinjection emf).

6.1. General Solution

After switching off the current, the distribution of holes in
the base is determined as a special case of Eq. (1.50) by substituting $j_0 = 0$ into that equation. Using the contractions of Eq. (1.51),
as well as the relationship (1.52), we obtain

$$p(X, \mathcal{T}) = p_1 \psi_4(X, \mathcal{T}). \qquad (2.45)$$

Since the voltage across the p-n junction is governed solely by the
impressed density of holes and is independent of the hole distribution in the base, we shall consider only the value of $p(0, \mathcal{T})$, which
will be denoted by $P(\mathcal{T})$.

Assuming that $X = 0$ in Eq. (2.45) and bearing in mind that in
deriving this equation we have neglected the value of p_{n0} compared
with $p(X, \mathcal{T})$, we obtain

$$p(\mathcal{T}) - p_{n0} = (p_1 - p_{n0}) \operatorname{erfc} \sqrt{\mathcal{T}}. \qquad (2.46)$$

In the Boltzmann statistics case, the voltage across the p-n junction is

$$u_{p-n} = \frac{kT}{q} \ln \frac{p_{x=0}}{p_{n0}}. \qquad (2.47)$$

(Later in this section, we shall use u instead of u_{p-n}, bearing in
mind that u represents the voltage across the p-n junction.)

Hence, using Eq. (1.21), we obtain,

$$u(\mathcal{J}) = \frac{kT}{q} \ln\left(1 + \left\{\exp\left[\frac{qu(0)}{kT}\right] - 1\right\} \operatorname{erfc} \sqrt{\mathcal{J}}\right), \qquad (2.48)$$

where u(0) is the voltage across the p–n junction when a forward current if is flowing through the junction (it is also the voltage at the moment of switching off the current). For all forward bias voltages of practical interest, we can ignore the unity in the braces in Eq. (2.48), compared with the exponential term. Then, for those values of time which satisfy $\operatorname{erfc} \sqrt{\mathcal{J}} \gg \exp\left[-\frac{qu(0)}{kT}\right]$, we obtain

$$u(\mathcal{J}) - u(0) = \frac{kT}{q} \ln \operatorname{erfc} \sqrt{\mathcal{J}}. \qquad (2.49)$$

Hence

$$\frac{d\left[u(\mathcal{J}) - u(0)\right]}{d\mathcal{J}} = -\frac{kT}{q} \frac{e^{-\mathcal{J}}}{\sqrt{\pi \mathcal{J}} \operatorname{erfc} \sqrt{\mathcal{J}}}, \qquad (2.50)$$

which gives the following expression for large values of $\mathcal{J}$ when the expansion (1.57) of the error function is used:

$$\frac{d\left[u(\mathcal{J}) - u(0)\right]}{d\mathcal{J}} = -\frac{kT}{q}\left(1 + \frac{1}{2\mathcal{J}} - \frac{3}{4\mathcal{J}^2} + \cdots\right). \qquad (2.51)$$

Restricting ourselves to the zeroth-order approximation, we find that, in a certain range of values of time, the decay of the post-injection voltage is linear:

$$\frac{du(t)}{dt} = -\frac{kT}{q} \frac{1}{\tau_p}. \qquad (2.52)$$

This expression has been obtained by Gossick [4, 10, 52, 53] Leher-handler and Giacoletto [54], and by Henderson and Tillman [30]. We shall now determine the range of values of $\mathcal{J}$ in which Eq. (2.52) is valid. At low values of $\mathcal{J}$, the neglecting of all terms containing $\mathcal{J}$ in Eq. (2.51) gives an error smaller than 20% provided $\mathcal{J} > 2$. For large values of $\mathcal{J}$, Eq. (2.49) is valid when

$$\exp\left[\frac{qu(0)}{kT}\right] \cdot \operatorname{erfc} \sqrt{\mathcal{J}} \gtrsim 5. \qquad (2.53)$$

After suitable simplifications, we find that Eq. (2.52) is correct to within 20% in the range

$$2 < \mathcal{J} < \frac{qu(0)}{kT} - (2 - 4).$$

$$(2.54)$$

Since the value of $qu(0)/kT$ may reach 10–15 for germanium diodes and 20–25 for silicon diodes, it follows from Eq. (2.54) that the postinjection emf decays linearly in a fairly wide time interval exceeding $10\tau_p$.

We can estimate, by analogy with Eq. (2.54), the values of the voltage across the p–n junction for which the decay of this voltage is linear. From the form of Eq. (2.48), it is evident that this equation is satisfied when $u(t) > kT/q$. Thus, the linear part of the function $u(t)$ extends to values of $\mathcal{J}$ defined by Eq. (2.54) and to values of $u(t)$ close to kT/q.

At very low values of $u(t)$, when the second term in the logarithm of Eq. (2.48) becomes smaller than the first term and consequently $u(t) \ll kT/q$, the expression (2.48) can be simplified. After expanding the error function and the logarithm as series, we obtain

$$u(t) \simeq \frac{kT}{q} \left[e^{\frac{qu(0)}{kT}} - 1 \right] \frac{e^{-\mathcal{J}}}{\sqrt{\pi \mathcal{J}}}.$$

$$(2.55)$$

Since the value of $\mathcal{J}$ is in this case very large $(\mathcal{J} \gg 1)$, we may assume that the decay of the potential across the p–n junction during the second (reverse) phase of the transient is exponential. This result was obtained by Gossick [52] on the basis of approximate calculations.

The expression (2.52) can be obtained also from simpler considerations. If we assume that, when the forward current is switched off, the decay of the excess hole density near the p–n junction is solely due to recombination i.e., if we neglect the diffusion from the p–n junction to the base interior), we find that in the case of a linear recombination law the definition of the lifetime yields

$$p(t) = (p_1 - p_{n0}) e^{-t/\tau_p}.$$

Substituting this expression into Eq. (2.47) and using Eq. (1.21), we immediately obtain Eq. (2.52).

Thus, after a rapid initial leveling of the hole distribution in the contact region [in a time interval $(1-2)\tau_p$], which is due to recombination and diffusion, the subsequent decay of the impressed value of the hole density is almost wholly due to recombination. An estimate obtained using Eq. (2.45) shows that, for example, when $\mathcal{J} = 2$ the difference between the hole densities at points $X = 0$ and $X = 0.5$ does not exceed 2%, i.e., there is practically no diffusion.

The difference between the exact solution (2.48) and the approximate solution (2.55), in which an exponential hole density decay is assumed, is shown in Fig. 2.19 where the curves are plotted for the case when $qu(0)/kT = 8$. We can see that practically throughout the transient process the exact value of $u(t)$ differs considerably from the value given by the approximate formula. However, the slopes of the approximate and exact dependences are practically the same (within 15%) at all points in the range $2 < \dot{\mathcal{J}} < 6$; this confirms the validity of the approximation (2.52) and the correctness of our definition of the range of values of $\mathcal{J}$ given by Eq. (2.54) and obtained on the assumption that the voltage decays linearly.

6.2 Allowance for Carrier Accumulation in the Emitter Region

A generalization of Eq. (2.48), which is improtant in experimental investigation, has been deduced by North. He has solved

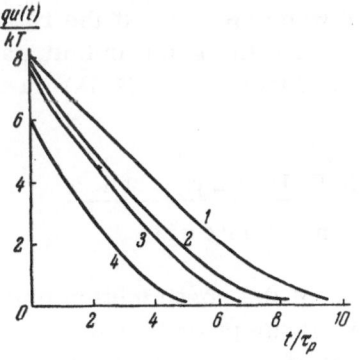

Fig. 2.19. Decay of the postinjection emf, $qu(\tau)/kT$, in accordance with the approximate (1) and exact (2) formulas, obtained assuming an infinitely long flow of the forward current ($t_f = \infty$), and in accordance with the approximate formula for $t_f = \tau_p$ (3) and $\tau_f = 0.1\tau_p$ (4).

the problem of the decay of the postinjection voltage for the case
when the forward current injects minority carriers into the p- and
n-type regions (i.e., the injection efficiency is $\gamma \neq 1$). In this case,
holes accumulate in the n-type region and electrons in the p-type
region; the decay of the voltage across the p-n junction is govered
by two lifetimes: τ_p and τ_n. Integration of the diffusion equation
gives

$$\frac{\exp\left[\dfrac{qu(t)}{kT}\right]-1}{\exp\left[\dfrac{qu(0)}{kT}\right]-1} = \frac{j_{sp}}{j_{sp}-j_{sn}}\,\mathrm{erfc}\,\sqrt{\frac{t}{\tau_p}} -$$

$$-\frac{j_{sn}}{j_{sp}-j_{sn}}\,\mathrm{erfc}\,\sqrt{\frac{t}{\tau_n}} + \frac{\sqrt{j_{sp}j_{sn}}}{j_{sp}-j_{sn}}\,\sqrt{B_1}\,\times$$

$$\times\, e^{-B_2 t}\left(\mathrm{erf}\,\sqrt{\frac{j_{sn}}{j_{sp}}\,B_1\,\frac{t}{\tau_p}} - \mathrm{erf}\,\sqrt{\frac{j_{sp}}{j_{sn}}\,B_1\,\frac{t}{\tau_p}}\right), \qquad (2.56)$$

where $j_{sp} = p_{n0}qD_p/L_p$ and $j_{sn} = n_{p0}qD_n/L_n$ are the densities of the
saturation currents for the minority carriers in the n- and p-type
regions, respectively; the coefficients B_1 and B_2 are:

$$B_1 = \frac{j_{sp}j_{sn}(\tau_p-\tau_n)}{j_{sp}^2\tau_p - j_{sn}^2\tau_n}, \qquad B_2 = \frac{j_{sp}^2 - j_{sn}^2}{j_{sp}^2\tau_p - j_{sn}^2\tau_n}.$$

The expression (2.56) should be used when $\gamma \neq 1$ and when the
accumulation of electrons in the p-type region cannot be ignored
(in particular, it should be used for some types of diffused diode,
for p-n junctions prepared by growing single crystals from the
melt, and for epitaxial structures).

The expression (2.48) is a special case of Eq. (2.56): this
can be seen from the fact that $j_{sn} \rightarrow 0$ when $n_{p0} \rightarrow 0$. If the rates
of recombination of electrons and holes are the same in both parts
of the diode ($\tau_p = \tau_n$), then Eq. (2.56) reduces to Eq. (2.48) irrespective of the values of j_{sp} and j_{sn}.

6.3. Postinjection EMF Decay with a
Finite Load in the Circuit

All the expressions for the decay of the postinjection emf
obtained so far have been based on the assumption that $R_l = \infty$.
However, in practice, the measurement of the voltage across a

diode always involves the connection of a parallel finite load (equal
to the input resistance of the measuring instrument). Very considerable deviations from the condition $R_l = \infty$ are observed in the
measurement of time intervals in the nanosecond range since in
this case the load resistances are coaxial cables with characteristic impedances not exceeding several hundred ohms. In some cases,
a capacitance is connected in parallel with the diode. Thus, at the
end of a forward current pulse, the diode is short-circuited either
across a finite resistance R_l or across a capacitor C but, as in the
open-circuit case, there is still no external voltage (Fig. 2.20).
From the physical point of view, it is obvious that the lower the
value of R_l and the higher the value of C, the stronger is the departure of the switching conditions from the open-circuit case
$(R_l = \infty)$.

The presence of a resistance R_l in a circuit means that when
the forward current stops flowing through a diode, a reverse current begins to flow due to the intrinsic emf of the p-n junction which
is equal to $u(t)/R_l$. The flow of such a transient reverse current
accelerates the loss of holes from the base and the role of this current increases with decrease of the value of R_l.

A capacitor connected in parallel with a diode become charged,
during the flow of the forward current, to a voltage $u(0)$; when the
external current generator is disconnected, a forward current still
flows through the diode and this current is fed by the charge stored
in the capacitor. This will be observed until the whole capacitor
charge flows into the base of the diode, where it gradually recombines. The higher the value of the capacitance C, the longer the

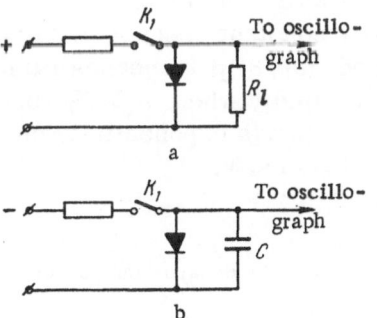

Fig. 2.20. Circuits for observing the
postinjection emf decay in a diode
connected to a load resistance (a) or
a capacitance (b). The switch K_1 is
opened at a time $t = 0$.

period that this capacitance can maintain the forward current and the more important its influence on the transient process.

Thus, the presence of a finite resistance in the diode discharge circuit accelerates the decay of the postinjection emf and the presence of a capacitor slows down this decay.

We shall now obtain a quantitative estimate of the transient process in a diode switched as shown in Figs. 2.20a or 2.20b.

The exact calculation of the postinjection emf in a diode connected to a resistance requires integration of Eq. (1.13) with a nonlinear boundary condition, which can be carried out only numerically.

However, approximate quantitative relationships can be obtained by making some simplifying assumptions.*

We shall assume that the influence of the load resistance on the form of the function u(t) is negligible until the loss of the excess charge in the interior of the base due to recombination exceeds the loss due to the current flowing through the load resistance R_l under the action of the emf u(t) acting across the p-n junction. This conclusion follows from the fact that the emf across the p-n junction (which governs the current through R_l) and the hole density in the base (which governs the intensity of recombination) are related by an exponential law. Let us assume that at some value $J = J_l$ the following relationship is satisfied:

$$\left(\frac{\partial p}{\partial t}\right)_j = \left(\frac{\partial p}{\partial t}\right)_R, \tag{2.57}$$

where the subscripts j and R represent the current and recombination loss of holes. It is obvious that at all values of J close to J_l, the value of $(\partial p/\partial t)_j$ is practically constant because u(t) depends linearly on t. On the other hand, $(\partial p/\partial t)_R$ increases rapidly when $J < J_l$ and decreases equally rapidly when $J > J_l$ because the density of holes in the base depends exponentially on time and $(\partial p/\partial t)_R$ is proportional to the hole density.

---

* An estimate of the influence of the load resistance on the decay of the postinjection emf in a planar diode has been obtained by O. K. Mokeev.

Thus, our assumption means that up to the moment $\mathcal{J} = \mathcal{J}_l$ the decay of the post injection emf proceeds in the same way as under the open-circuit conditions (i.e., when $R_l = \infty$) but beginning from this moment the decay of this emf is governed solely by the load resistance.

Using the expression for the total charge of holes stored in the base, given by Eq. (2.8) and assuming a linear recombination law, we obtain

$$\left(\frac{\partial p}{\partial t}\right)_R = \frac{1}{\tau_p} \frac{i_f \tau_p}{q} e^{-t/\tau_p}. \tag{2.58}$$

Dividing each term in Eq. (2.48) by qR_l, we obtain

$$\left(\frac{\partial p}{\partial t}\right)_j = \frac{kT}{q^2 R_l} \ln \operatorname{erfc} \sqrt{\mathcal{J}} + \frac{u(0)}{qR_l}. \tag{2.59}$$

Equating these two expression at $\mathcal{J} = \mathcal{J}_l$ and making the necessary transformations, we get the following equation for the determination of $\mathcal{J}_l$:

$$\ln \operatorname{erfc} \sqrt{\mathcal{J}_l} + \frac{qu(0)}{kT} = \frac{q}{kT} i_f R_l e^{-\mathcal{J}_l}. \tag{2.60}$$

Figure 2.21 shows graphically the values of $\mathcal{J}_l$ found by solving this equation for u(0) = 0.4 and 0.7 V, which are typical values for planar silicon diodes.

In accordance with the assumption made earlier, the loss of holes from the moment $\mathcal{J} = \mathcal{J}_l$ is solely due to the current flowing through the resistance R_l; in the first approximation, we may assume that the value of this current decreases exponentially with time in accordnce with the law

$$i(t) = i(t_l) \exp(-t/\tau_x), \tag{2.61}$$

where τ_x is an unknown time constant. Equating the charge retained in the base at a time t_l, which is $Q_{st}\exp(-t_l/\tau_p)$, to the charge carried by the reverse current in accordance with Eq. (2.61), and finding the value of $i(t_l)$ from Eq. (2.58), we obtain $\tau_x = \tau_p$.

It follows that the postinjection emf in a diode connected to a resistance decreases at first linearly (as in the open-circuit case) but from a moment $\mathcal{J}_l$, defined by Eq. (2.60), the emf decreases exponentially with a characteristic time equal to τ_p.

Let us now find the value of the resistance in the diode circuit which can be regarded as practically infinite by defining it as the value $(R_l)_\infty$ at which the form of the function u(t) does not depart from linearity. Assuming that $\mathcal{J}_l = \mathcal{J}_{max}$ [where $\mathcal{J}_{max}$ is the limit of the linear part of u(t)] and using Eq. (2.53), we find from Eq. (2.60) that

$$R_{l\infty} \simeq 1.5\frac{kT}{qi_f}\, e^{\mathcal{J}_{max}} \simeq R_i \cdot e^{\mathcal{J}_{max}}. \qquad (2.62)$$

This equation is obtained neglecting a factor 1.5 and assuming that kT/qi_f represents the differential resistance of the p-n junction during the flow of a forward current i_f. The expression for R_i is obtained in an elementary manner by differentiating the current-voltage characteristic of the p-n junction:

$$i_f = i_s\big(e^{\,qu/kT} - 1\big). \qquad (2.63)$$

When, for example, the linear part of the postinjection emf decay extends to $6\tau_p$ (as in Fig. 2.19) for a forward current $i_f = 1$ mA,

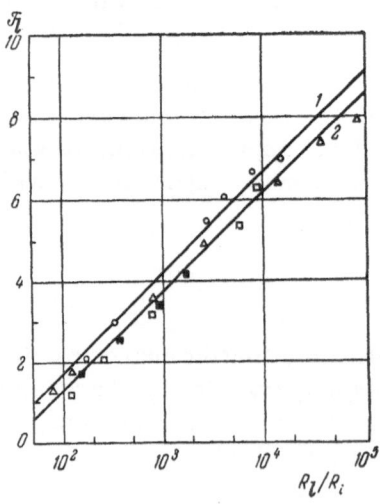

Fig. 2.21. Theoretical dependences of the duration of the linear part of the postinjection emf decay on R_l/R_i for u(0) = 0.4 V (1) and for u(0) = 0.7 V (2). Experimental points represent the results obtained for three diodes (for an explanation see end of § 9.4).

it follows that $\mathcal{J}_{max} = 6$, and $R_i = 25\Omega$. In this case, the circuit can be assumed to be open when $R_l \gtrsim 10$ kΩ.

Gossick [52] estimated the influence of a parallel capacitance on a transient in a diode. For a very small voltage u(0) he obtained

$$u\left(t\right) = \frac{kT}{q}\left(1 - e^{-\frac{qu\left(0\right)}{kT}}\right)e^{-\frac{qi_sC}{kT}t} \tag{2.64}$$

For a very large reverse voltage, he found

$$u\left(t\right) = u\left(0\right) - \frac{i_s t}{C}. \tag{2.65}$$

The last equation is self-evident: when a diode is blocked by the capacitor voltage, the capacitor charge decays solely by the flow of a reverse current, equal to the saturation current, through the diode.

6.4. Nonstationary Forward Current

Ko [27] obtained an expression for the decay of the postinjection emf when the forward current through a p-n junction before switching flows for a finite time t_f. The solution of the diffusion equation with the initial condition given by Eq. (1.55) and the boundary condition

$$\left(\frac{\partial p}{\partial X}\right)_{X=0} = 0, \tag{2.66}$$

corresponding to the case $R_l = \infty$, yields the following expression

$$u\left(\mathcal{J}\right) = \frac{kT}{q}\ln\left[1 + \frac{i_f}{i_s}\left(\text{erf}\sqrt{\mathcal{J}_f + \mathcal{J}} - \text{erf}\sqrt{\mathcal{J}}\right)\right]. \tag{2.67}$$

where $\mathcal{J}$ is measured (as usual) from the end of the forward current pulse. Neglecting the same quantities as in the derivation of Eq. (2.49) we obtain

$$u\left(\mathcal{J}\right) = u\left(0\right) + \frac{kT}{q}\ln\left(\text{erf}\sqrt{\mathcal{J}_f + \mathcal{J}} - \text{erf}\sqrt{\mathcal{J}}\right). \tag{2.68}$$

We shall now analyze this expression. The value of the voltage across the p-n junction at $\mathcal{J} = 0$, when the decay of the postinjection emf begins, is not equal to u(0) but is always less than this quantity and is governed by the duration of the forward current

$$u_{\mathcal{J}=0} = u(0) + \frac{kT}{q} \ln \operatorname{erf} \sqrt{\mathcal{J}_f} \qquad \text{when } \mathcal{J}_f \neq 0. \qquad (2.69)$$

In the nonstationary forward current case, we shall denote the initial voltage by $u_{\mathcal{J}=0}$, and use $u(0) = \frac{kT}{q} \ln \frac{p_1}{p_{n0}}$ to represent that voltage which would have been established across the p-n junction if the forward current had flowed for an infinitely long time.

Let us assume that $\mathcal{J} \gg \mathcal{J}_f$ and that $\mathcal{J} > 1$ [this is possible only when the values of $\mathcal{J}_f$ are not too small, because otherwise we cannot neglect the unity in the logarithm of Eq. (2.67)]. Then, differentiating Eq. (2.68), we easily obtain

$$\frac{du(\mathcal{J})}{d\mathcal{J}} = \frac{kT}{q} \frac{\dfrac{\exp(-\mathcal{J}-\mathcal{J}_f)}{\sqrt{\pi(\mathcal{J}+\mathcal{J}_f)}} - \dfrac{\exp(-\mathcal{J})}{\sqrt{\pi\mathcal{J}}}}{\operatorname{erf}\sqrt{\mathcal{J}_f+\mathcal{J}} - \operatorname{erf}\sqrt{\mathcal{J}}} \simeq -\frac{kT}{q}\left(1 + \frac{1}{2\mathcal{J}} + \cdots\right). \qquad (2.70)$$

Thus, in the case of switching from a nonstationery forward current, the decay of the postinjection emf takes place at the same rate as in the case $\mathcal{J}_f \to \infty$, which is demonstrated by the identity of Eqs. (2.70) and (2.51).

Graphical solutions of Eq. (2.68) are shown in Fig. 2.19 for several values of the parameter $\mathcal{J}_f$; they are obtained on the assumption that $qu(0)/kT = 0$. It is interesting to note that the initial value of the postinjection emf and the rate of its decay are not greatly affected when $\mathcal{J}_f$ is reduced. In practice, we can use the formula obtained for the case of a steady-state forward current for $\mathcal{J}_f \geq 1$. Similar estimates, obtained for switching conditions in which the linear decay region is longer, show that, in general, the forward current can be assumed to have reached a steady state when the duration of the flow of the forward current is $t_f \gtrsim (2-3)\tau_p$.

6.5 High Injection Levels

A calculation of the decay of the postinjection emf in the case of high injection levels, given in [55], is based on Eq. (1.18). The initial hole distribution is described by Eq. (2.43) and it is assumed that the injection level throughout the base is high ($\Delta \gg 1$) so that, consequently, $\alpha = \sqrt{2b/(b+1)}$. It is also assumed that the hole lifetime in the base is the same at all injection levels. This assumption is not in agreement with real conditions and it postulates that the driving effect on holes is solely due to the driving field observed when $\Delta \gg 1$. Solving the diffusion equation using these assumption and the boundary condition of the Eq. (2.66) type, i.e., assuming that $R_l \rightarrow \infty$, we obtain the self-evident relationship

$$\frac{du\,(t)}{dt} \simeq - \frac{kT}{q} \cdot \frac{1}{\tau_p} \cdot \frac{2b}{b+1}. \qquad (2.71)$$

In the derivation of Eq. (2.71), it is assumed, as before, that $\mathcal{J} \gg 1$ and $u(t) \gg kT/q$. The expression (2.71) is almost completely identical with Eq. (2.52): it differs from the latter only by a factor $2b/(b+1)$, which is ~ 1.4 for germanium and silicon at room temperature. Thus, the driving of holes by the field which appears at high injection levels reduces by a factor of $2b/(b+1)$ the time constant of the linear part of the postinjection emf decay. As $u(t)$ decreases so that the condition $\Delta < 1$ is reached, the time constant increases to a value equal to τ_p and Eq. (2.71) becomes identical with Eq. (2.52). Thus, the postinjection decay curve has a region corresponding to low injection levels, which is described by Eq. (2.52) even in the case of large initial currents through the diodes.

We must mention that oscillograms of the voltage observed when a large forward current is switched off in real diodes are of the form shown in Fig. 2.22. First, we observe a sudden fall of the voltage u_b across the base, which is followed by a more gradual decrease of the voltage. This behavior is explained by the appreciable importance of the ohmic voltage drop across the base in the case of large forward currents. When the current ceases to flow, this ohmic component of the voltage drop suddenly decreases to zero and we are left with the voltage drop across the p–n junction, which

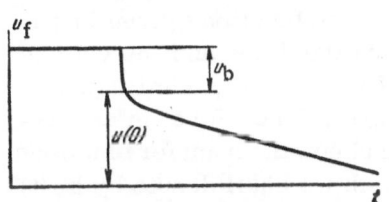

Fig. 2.22. Oscillogram of the decay of the voltage across a diode after the end of a forward current pulse of large amplitude [u_b is the voltage drop across the base, $u(0)$ is the voltage drop across the p–n junction].

is of the diffusion type and which decays relatively slowly with time. Thus, the value of u(0) is found from the total voltage drop across the diode after subtracting from it the ohmic drop u_b.

§ 7. SMALL-SIGNAL TRANSIENT CHARACTERISTICS OF A DIODE

In the preceding sections, we have considered the transient processes in a diode subjected to forward and reverse bias voltages whose amplitudes are sufficient to make the diode a strongly nonlinear element. In some radio-engineering applications, the interest lies in small-signal transients.

We shall consider the circuit shown in Fig. 1.1 and we shall assume that a steady-state forward current $i_f^{(0)}$ is flowing through the diode; at a time t = 0, a voltage step of $u_0 \ll kT/q$ is applied to the diode. We shall also postulate that all the assumptions made in §§ 2 and 3 (the diffusion approximation, the semi-infinte extent of the n-type region, $\gamma = 1$) still apply in this case.

The diode can now be replaced by a resistance whose value is equal to the differential resistance of the diode under the working conditions considered ($R_i = kT/qi_f^{(0)}$); during a transient process we must also take into account the effect of a diffusion capacitance, which shunts this resistance.

Thus, under small-signal conditions the circuit in Fig. 1.1 becomes linear and the definition of transfer functions given in § 1 applies to this case. The current through the diode and the voltage across it can be assumed to be known if we know the admittance of the circuit Y and the bias ratio $K = u_{p-n}/u_0$. Therefore, in an analysis of any transient processes we must know the transfer functions of the admittance and the bias ratio, which we shall denote by $h_Y(t)$ and $h_K(t)$, respectively.

A theoretical analysis of the general case has been carried out by Adirovich [5], and of some special cases, by Gossick [4]. For an arbitrary ratio of the resistances R_l and R_i, the process of establishment of the current in the circuit under the action of a single voltage step applied to the imput, is described by the expression

$$h_Y(\mathcal{J}) = \frac{1}{R_l + R_i}\left[1 + \frac{1}{a-1}\,\text{erfc}\,\sqrt{\mathcal{J}} - \frac{1}{a(a-1)}\exp\left(\mathcal{J}\,\frac{a^2-1}{a^2}\right)\text{erfc}\,\frac{1}{a}\sqrt{\mathcal{J}}\right]$$

$$(2.72)$$

where $a = R_l/R_i$ and, as before $\mathcal{J} = i/\tau_p$.

Consequently, the process of the establishment of the voltage across the p–n junction is described by the formula

$$h_K(\mathcal{J}) = 1 - R_l h_Y(\mathcal{J}).$$

$$(2.73)$$

The functions $h_Y(\mathcal{J})$ and $h_K(\mathcal{J})$ are shown, for several values of the parameter a, in Fig. 2.23. The lower the value of the parameter a, the faster are the transient processes. The physical reason for this is that, when a voltage pulse ($a \ll 1$) is applied to the diode, the current density during the establishment of equilibrium is always higher than the steady-state current density; therefore, the accumulation of the necessary number of excess holes is more rapid than under current generator conditions ($a \gg 1$).

In fact, by differentiating Eq. (1.65), which describes the distribution of holes in the base when a positive voltage step is applied to the p–n junction, we obtain the following expression for the density of the forward current during the establishment of equilibrium:

$$j(\mathcal{J}) = j_f\left[\frac{e^{-\mathcal{J}}}{\sqrt{\pi\mathcal{J}}} + \text{erf}\,\sqrt{\mathcal{J}}\right],$$

$$(2.74)$$

where j_f is the steady-state forward current density when $\mathcal{J} \to \infty$. For short durations of the forward current ($\mathcal{J} < 1$), Eq. (2.74) becomes

$$j(\mathcal{J}) \simeq j_f\,\frac{1}{\sqrt{\pi\mathcal{J}}}.$$

$$(2.75)$$

The accumulation of the hole charge, stored in the diode base, is given by

$$Q_{st}(\mathcal{J}) = \int_0^t j(t)\,dt \sim \frac{2\sqrt{\mathcal{J}}}{\sqrt{\pi}}\,Q_{st}(\mathcal{J} \to \infty).$$

$$(2.76)$$

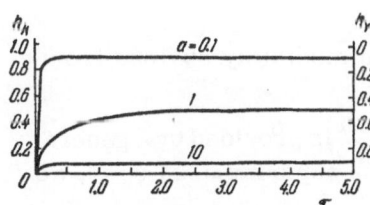

Fig. 2.23. Transfer function of the bias ratio and of the diode admittance for three values of a.

On the other hand, when a forward current pulse is applied to the diode, the accumulation of the hole charge in the base obeys the law

$$Q_{st}(\mathcal{J}) = Q_{st}(\mathcal{J} \to \infty)(1 - e^{-\mathcal{J}}) \simeq Q_{st}(\mathcal{J} \to \infty) \cdot \mathcal{J}, \qquad (2.77)$$

where the second equality is valid when $\mathcal{J} \ll 1$.

Comparison of the expressions (2.76) and (2.77) shows that the rate of accumulation of holes in the diode base immediately after switching (i.e., when $\mathcal{J} \ll 1$), is $1/\sqrt{\mathcal{J}}$ faster when a voltage pulse is applied to the p-n junction than in the case of application of a current pulse. Qualitatively, this conclusion follows from an analysis of the expression (2.74), in which $i(\mathcal{J}) > i_f$ for any value of $\mathcal{J}$.

We shall now derive analytic expressions for the transfer function in two limiting cases: the short-circuit condictions, i.e., when a pulse is applied from a voltage generator directly to the p-n junction ($R_l \ll R_i$); the open-circuit conditions, i.e., when the junction is connected to a current generator ($R_l \gg R_i$).

In the former case, when $\mathcal{J} \gg (R_l/R_i)^2$, we obtain

$$h_Y(\mathcal{J}) \simeq \frac{1}{R_i}\left(\frac{e^{-\mathcal{J}}}{\sqrt{\pi\mathcal{J}}} + \operatorname{erf}\sqrt{\mathcal{J}}\right), \qquad h_K(\mathcal{J}) \simeq 1. \qquad (2.78)$$

In the second case, we have

$$h_K(\mathcal{J}) \simeq \frac{R_i}{R_l}\operatorname{erf}\sqrt{\mathcal{J}}. \qquad (2.79)$$

We shall show that these transfer function are special cases of the corresponding transient response considered in preceding sections.

In the short-circuit case, this is immediately obvious from a comparison of Eqs. (2.78) and (2.74).

When a constant forward current $i_f^{(0)}$ is provided by a generator and a current pulse (equal to u_0/R_l) is superimposed on it, we obtain from Eq. (1.58):

$$p(\mathcal{J}) = \frac{u_0 L_p}{R_l S D_p q}\operatorname{erf}\sqrt{\mathcal{J}} + p_1^{(0)}, \qquad (2.80)$$

where $p_1^{(0)}$ is the impressed hole density corresponding to the current $i_f^{(0)}$ and defined in Eq. (1.25).

When a voltage step of amplitude u_0 is applied to the p-n junction, the voltage across the junction is given by

$$u(\mathcal{T}) - u_f^{(0)} = \frac{kT}{q} \ln \frac{p_1(\mathcal{T})}{p_1^{(0)}}.$$ (2.81)

Using the small-signal conditions ($u_0 \ll kT/q$) and expanding the logarithm into a series, we obtain

$$h_K(\mathcal{T}) = \frac{u(\mathcal{T}) - u_f^{(0)}}{v_0} = \frac{kT}{q} \frac{L_p}{R_l S D_p q p_1^{(0)}} \operatorname{erf} \sqrt{\mathcal{T}},$$ (2.82)

and when we substitute in it $\dfrac{kT}{q} \dfrac{L_p}{SqD_p p_1^{(0)}} = R_l$, we obtain Eq. (2.79).

§8. METHODS FOR THE OBSERVATION OF TRANSIENT PROCESSES IN DIODES

8.1 Measuring Apparatus

To observe a transient process in a diode, we need a generator, producing a switching pulse, and an oscillograph for the observation of the time dependence of the current through the diode, or of the voltage across it. Sometimes, an oscillograph is replaced by some other instrument which can measure a given parameter of the transient process. The apparatus used in studies of transient processes in diodes always consists of theses two units: a generator and a measuring device.

Circuits for the Observation of Transient Reverse Currents. The simplest circuit for the observation of the decay of a transient reverse current through a diode, which follows the switching from a steady forward state to a reverse voltage, is shown in Fig. 2.24 [29]. Whe the contacts of a mercury relay are open, a forward current flows through the investigated diode from a source E_1 and the value of this current is governed by the battery voltage E_1 and a current-limiting resistance R_1 (the value of this resistance is usually 1-2 kΩ, which is considerably higher than the forward resistance of the diode and the

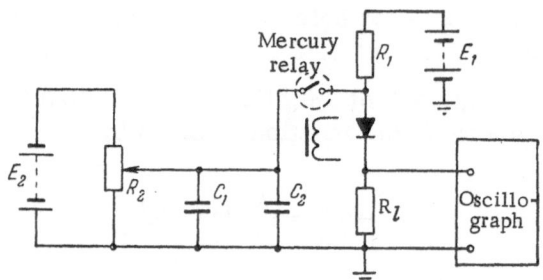

Fig. 2.24. Circuit for the observation of a transient
reverse current through a diode switched from the
forward to the reverse voltage

load resistance R_l). When the relay contacts are closed, a reverse
voltage is applied to the diode by E_2 and this voltage is equal to the
difference between the potentials across capacitors C_1 and C_2 before switching. The high value of the resistance R_1 effectively suppresses the influence of the source E_1 on the diode after switching.
The voltage across a resistance R_2, which is always proportional
to the reverse current, is viewed on the screen of a fast-response
cathode-ray oscillograph. The capacitances C_1 and C_2, connected
in parallel, are made sufficiently large so that we can assume that
the voltage generator has a zero output resistance at the moment of
switching. The lower the value of R_l , the closer are the conditions
in the diode circuit (after switching) to the short-circuit case.
Usually, the load resistance ranges from several ohms to tens of
ohms. If the value of R_l is reduced below these values, the current
sensitivity of this circuit is reduced but reduction of the load resistance has practically no influence on the transient process because
the ohmic resistance of the diode base interior is usually several
ohms. It we assume that R_l = 10 Ω, then we find that the maximum
value of the reverse current is $i_0 \approx 1$ A when $U_r = 10$ V. If the value of the forward current is $i_f \approx$ 10–50 mA (which is typical of the
investigated diodes), then in all cases we have B = i_0/i_f > 20 which
is (§ 5) practically indistinguishable from the case B = ∞ (i.e., R_l = 0).

The correct selection of the value of the capacitance C_1 is
best made on the basis of the relationships between the charges:
the capacitor charge should be at least two orders of magnitude
larger than the recovered charge of the diode. In this case, the
reverse voltage across the diode will be constant throughout the

transient process. Let us assume, for example, that the investi-
gated diode has $\tau_p \approx 10\ \mu\text{sec}$ and that the measurements are carried
out during the switching from a forward current $i_f = 10$ mA to a reverse voltage $U_r = 10$ V. Then, we must satisfy the relationship
$Q_C \gtrsim 100\ Q_{rec}$ and hence $C_1 \gtrsim 5\mu\text{F}$. The capacitance C_2 is usually
an order of magnitude smaller than C_1; the second capacitance is ne-
cessary to suppress the distortion of the front of the reverse pulse.
Sometimes, the capacitors C_1 and C_2 in the circuit of Fig. 2.24 are replaced with a small resistance (of the order of several ohms) which
acts as the ouput resistance of the voltage generator.

When a mercury relay is used, the switching time of a diode
from one state to another may be shorter than 10^{-9} sec, which — in
the majority of cases — is much shorter than the hole lifetime in
the diode base. Since the switching frequency of a mercury relay
is low (of the order of $10-10^2$ cps), the forward–current and reverse–current and reverse–voltage conditions reach a steady state
during the time between two consecutive pulses.

A typical oscillogram of a reverse current transient (obtained
using the circuit in Fig. 2.24) is shown in Fig. 2.25. We can easily
see that time intervals can be measured satisfactorily only when
the reverse current is not less than a tenth of i_f.

Methods for Increasing the Current Sensitivity.
When it is necessary to investivate transient processes in the range
of very low instantaneous values of the reverse current, the sensitivity of the measuring circuit can be increased by increasing the
load resistance R_l. This increase of
the resistance does not affect the
tail part of the transient process
(this follows from the § 5 and theoretical curves in Fig. 2.14). To reduce the voltage drop across the
load resistance during the passage
of the forward current, it is shunted
by a series of vacuum-tubes or fast-
response semiconductor diodes.
A circuit with a high current sensitivity is shown in Fig. 2.26 [56].
The inclusion of a diode D reduces

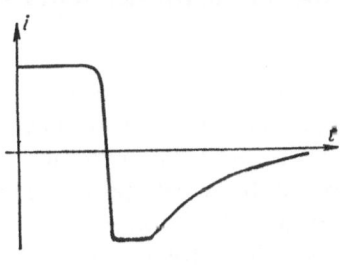

Fig. 2.25. Oscillogram of the current during switching

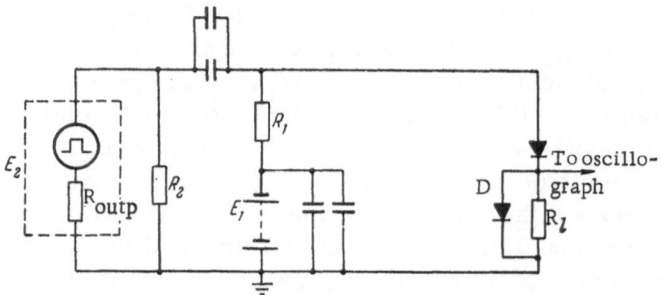

Fig. 2.26. Circuit for the observation of a transient reverse current through a diode in the case of high values of the load resistance. $R_1 = 1000\,\Omega$; $R_2 = 600\,\Omega$; $R_l = 2000\,\Omega$; $R_{outp} = 600\,\Omega$.

the total signal across R_l, recorded by means of an oscillograph, and consequently the signal can be amplified without any danger of overloading the oscillograph amplifier and of causing nonlinear distortions. It is interesting that in the circuit of Fig. 2.26 the output resistance of the voltage generator may be considerably higher than that in the circuit of Fig. 2.24. This does not distort the transient process because the value of R_l is high.

One of the modifications of the circuit in Fig. 2.26 uses sources E_1 and E_2 with reversed polarities [57]. In this case, a diode is always blocked by the voltage of the source E_1 and the forward current flows through it only in the form of relatively short pulses. Such connection of the diode is necessary when transient processes are investigated at high values of the forward current which would damage or change irreversibly the properties of the diode if it were allowed to flow continously.

The next logical development of the method considered here is the circuit shown in Fig. 2.27 [58]. Here, another diode is connected in series with the load resistance and in opposition to the shunting diodes; this separates completely the forward and reverse currents through the diode under investigation. The same circuit can be used to measure the recovered charge if the load resistance is replaced by a microammeter shunted by a capacitance [59]. In this case, when the repetition frequency of the reverse pulses f is known and the average value of the transient reverse current i_a is measured with the microammeter, we can easily find the recovered

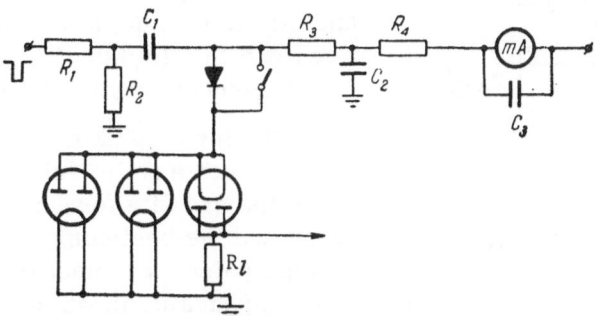

Fig. 2.27. Circuit for the observation of switching oscillograms in which the forward and reverse currents are separated.
$R_1 = 33\ \Omega$; $R_2 = 18\ \Omega$; $R_l = 100\ \Omega$; $C_1 = 0.1\ \mu F$; $C_2 = C_3 =$
0.01 μF.

charge using the self-evident relationship:

$$Q_{\text{rec}} = i_a f^{-1}. \tag{2.83}$$

Other variants of the circuits used in the laboratory investigations
of transient processes in diodes are discussed in a review by the
present author [40].

Various methods of increasing the sensitivity of the basic
circuit in Fig. 2.24 make it possible to measure time intervals even
when the current i(t) is of the order of 10^{-4} A. However, this value
is considerably higher than the steady-state reverse current, which
is close to the saturation current.

Further decrease of the measurable reverse currents can be
achieved by using a stroboscopic method, which was first employed
first by Henderson and Tillman [30] to investigate transient processes in diodes. The essence of this method is that only a small
part, corresponding to the time interval of interest, of the total
transient reverse current pulse is used; this part is then amplified
independently of the rest of the pulse. If this operation is repeated
for successive time intervals, it is possible to achieve a very high
current sensitivity in the investigation of the transient response.
A detailed description of this method is given in a specialized mo-
nograph [60]; applications to semiconductor devices are described
in [61-67]. Henderson and Tillman [30] were able to measure the

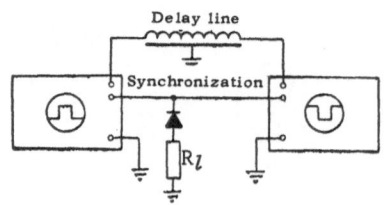

Fig. 2.28. Circuit for the investigation of transient processes in a diode in the case of a delay before the application of a reverse current pulse.

values of i(t), equal to 0.001 i_f, and others [63, 64] have been able to extend this to 0.0001 i_f in the reverse current range 5–10 μA.

To investigate transient processes when a finite delay separates the end of a forward current pulse and the beginning of reverse voltage pulse, it is necessary to to use apparatus in which both the forward and reverse voltages are in the form of pulses whose positions on the time scale can be varied. One of the suitable circuits is shown in Fig. 2.28 (its operation is self-evident).

Measurement of the Recovery Phase Duration and the Postinjection Voltage Tail. To measure the duration t_1 of the first (recovery or high–conductance) phase after switching, we can use the circuit in Fig. 2.24 with a suitably increased load resistance R_l or one of the modifications of this circuit. Some experimental difficulties are encountered in the exact determination of the end of the recovery phase. At this moment, the voltage begins to rise rapidly across the diode and this produces stray peaks due to the inductance of the connecting leads. The effect is particularly strong when the switching currents i_f and i_0 are high. Thus, in an investigation of transient processes in diodes with $i_0 \approx$ 1–10 A [51] it has been necessary to place the diode in a cylindrical chamber in order to reduce these stray peaks. In this case, the diode acts as the central core of a coaxial line. The load

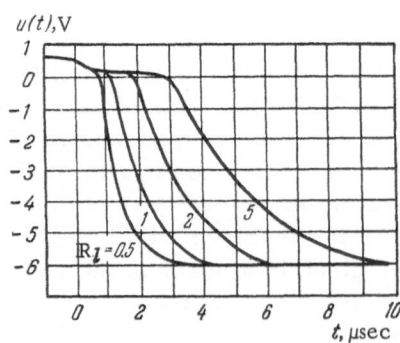

Fig. 2.29. Oscillograms showing the establishment of the reverse voltage across a diode when it is switched in the circuit shown in Fig. 2.24; the various values of R_l are given in k Ω.

resistance, R_l , in the form of a Manganin wire, also serves as the core of a coaxial cable connecting a pulse generator with the diode chamber. The leads from the diode chamber and the load resistance are also in the form of coaxial cables.

In the determination of the duration of the recovery phase we can investigate, instead of the reverse current transient, the time dependence of the voltage across the diode. To do this, the diode and the load resistance R_l in the circuit of Fig. 2.24 have to be interchanged. Typical oscillograms, showing the time dependence of the reverse voltage for a germanium diode, are given in Fig. 2.29. The moment of transition of the voltage through the zero value (corresponding to a sharp bend in the transient characteristic) effectively determines the end of the recovery phase.

To observe the decay of the postinjection emf, Gossick [52] used the circuit shown in Fig. 2.30b. Figure 2.30a shows a pulse generator circuit in which a segment of a long line is used. By varying the value of R_l in the circuit, we can investigate its influence on the form of the transient process. However, it is difficult to use this circuit under the open-circuit conditions because, at high values of R_l, it is necessary to apply large-amplitude voltage pulses to the diode circuit in order to obtain the required value of the forward current. It is much more convenient to use the circuit shown in Fig. 2.31, suggested by Lederhandler and Giacoletto [54], for the investigation of p-n junctions in alloyed transistors. Here, a vacuum-tube or a fast-response semiconductor diode breaks the circuit after stopping the flow of the forward current and therefore the voltage drop across the diode takes place in the absence of a current through it. In practice, it is necessary to connect a battery providing a voltage of about 1 V between the pulse generator out-

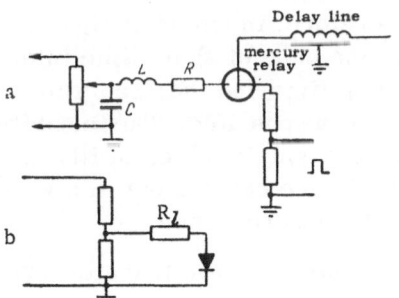

Fig. 2.30. a) Pulse generator circuit and b) circuit for the observation of the postinjection emf decay.

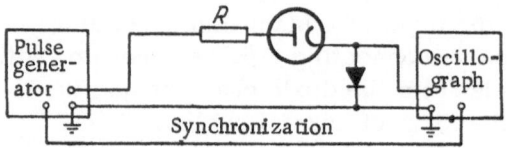

Fig. 2.31. Circuit for the observation of the post-injection decay of the voltage across a diode under open-circuit conditions.

put and the anode of the vacuum tube; this ensures that the vacuum diode is slightly blocked at the end of the forward current pulse. This is necessary to prevent the flow (through this vacuum diode) of the residual current, which may reach several tens of microamperes. Experimental investigations have shown that the nature of the transient process is practically independent of the pulse duration (provided it is greater than a certain minimum value) or the pulse repetition frequency. The pulse amplitude and the output impedance of the pulse generator are also unimportant. The important characteristics of the circuit are the resolution time of the oscillograph, the linearity of its horizontal and vertical amplification, the sensitivity, and the decay time of the current pulse.

It is very difficult to use rectangular test pulses in investigations of transient processes in diodes in the subnanosecond range. Because of this, sinusoidal signals are widely used in the subnanosecond range. The corresponding formulas must be altered accordingly in the interpretation of the experimental results.

8.2. Experimental Errors

The foregoing limiting factors are common to all the measuring circuits considered so far. Therefore, in any investigation of transient processes, particularly in the case of short time intervals, we should try to make the switching time and the resolution time of the measuring instrument as short as possible. The linearity of the amplification is important only when small values of $i(t)$ or $u(t)$ are measured or when measurements are carried out in a wide range of instantaneous values of the transient response.

In those cases when the measured duration of a transient process is comparable with the switching time of the measuring circuit, it is necessary to estimate the values of the possible error.

Such estimates have been obtained by Il'enkov [66] and Éidukas [67] for linear and exponential fronts of switching pulses.

Let us consider the case of a linear variation of the voltage at the output of the circuit shown in Fig. 1.1 during switching [68]. The dispersal of the stored charge should take place in the same way as in the case of instantaneous switching but the process should be somewhat slower since, during a gradually rising reverse voltage, the removal of holes from the base by the transient reverse current will be slower than in the case of a vertical voltage step.

We shall find first the distribution of the density of holes injected into the base for a linearly increasing forward current and we shall integrate the diffusion equation (1.13) for zero initial conditions and a boundary condition of the type

$$\left(\frac{\partial p}{\partial X}\right)_{X=0} = -\alpha \mathcal{J},\qquad (2.84)$$

where the coefficient α represents the steepness of the rising front of the forward current. The solution can be obtained using a two-dimensional Laplace–Carson transformation (cf. §3) and this solution is of the form

$$p(X, \mathcal{J}) = \alpha\left\{\frac{2\mathcal{J}-1}{4}\left[e^{-X}\operatorname{erfc}\left(\frac{X}{2\sqrt{\mathcal{J}}}-\sqrt{\mathcal{J}}\right)-e^{X}\operatorname{erfc}\left(\frac{X}{2\sqrt{\mathcal{J}}}+\sqrt{\mathcal{J}}\right)\right]-\right.$$

$$-\frac{X}{4}\left[e^{-X}\operatorname{erfc}\left(\frac{X}{2\sqrt{\mathcal{J}}}-\sqrt{\mathcal{J}}\right)+e^{X}\operatorname{erfc}\left(\frac{X}{2\sqrt{\mathcal{J}}}+\sqrt{\mathcal{J}}\right)\right]$$

$$\left.+\frac{\sqrt{\mathcal{J}}}{\sqrt{\pi}}e^{-\frac{X}{4\mathcal{J}}-\mathcal{J}}\right\} = \alpha F(X, \mathcal{J}).\qquad (2.85)$$

We shall assume that a constant forward current i_f has been flowing for an infinitely long time through a diode so that a steady–state distribution is established in the base. At some moment, the diode is switched to the reverse voltage by a pulse with a linear front whose characteristic parameters are shown in Fig. 2.32. Like the quantity $\mathcal{J}$, the time parameters of the pulse (a and b) are expressed in dimensionless units, i.e., they represent time divided by τ_p.

Since the diode circuit practically always contains a load resistance, which limits the reverse current, we shall consider a

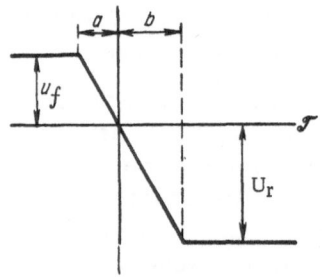

Fig. 2.32. Time dependence of the input voltage during the switching of a diode from the forward to reverse voltage by a pulse with a linearly rising front.

circuit with a load resistance. During the initial phase of the transient process, until the hole density near the p–n junction decreases to zero, the reverse current depends on the applied voltage. We shall denote the duration of this phase by $\mathcal{J}'_1$, by analogy with duration of the recovery phase. We are particularly interested in the case when the first phase ends before the reverse voltage reaches its maximum value. During the rise time, the reverse pulse and as long as $p(0, \mathcal{J}) > 0$, the hole distribution can be obtained by superimposing several functions of the type given by Eq. (2.85). This hole density distribution is of the form

$$p(X,\ \mathcal{J}) = p_1 \left[e^{-X} - \frac{1}{a} F(X,\ \mathcal{J} + a) - \frac{aB - b}{ab} F(X,\ \mathcal{J}) \right],\quad (2.86)$$

where p_1 and B are the limiting values, i.e., p_1 represents the impressed hole density in the case of a steady–state forward current of density j_f [this value of p_1 is given by Eq. (1.25)], and B = j_0/j_f, where $j_0 = U_r/SR_l$ (cf. Fig. 2.32).

To determine $\mathcal{J}'_1$, we shall equate to zero the value of p(0, $\mathcal{J}$) in Eq. (2.86). After self–evident transformations, we obtain the following expression which can be used to find $\mathcal{F}_1$:

$$\frac{(2\mathcal{J}'_1 + a) - 1}{2a}\ \mathrm{erf}\ \sqrt{\mathcal{J}'_1 + a} + \frac{\sqrt{\mathcal{J}'_1 + a}}{a\sqrt{\pi}}\ e^{-(\mathcal{J}'_1 + a)} +$$

$$+ \frac{aB - b}{b} \left(\frac{2\mathcal{J}'_1 - 1}{2a}\ \mathrm{erf}\ \sqrt{\mathcal{J}'_1} + \frac{\sqrt{\mathcal{J}'_1}}{a\sqrt{\pi}}\ e^{-\mathcal{J}'_1} \right) = 1. \quad (2.87)$$

Right up to the moment $\mathcal{J}'_1$, the rise of the transient reverse current through the diode follows the rise of the reverse voltage, i.e., it is described by a linear function. At the moment $\mathcal{J}'_1$ the current i(t) reaches its maximum value and than begins to decrease monotonically since the reverse resistance of the p–n junction begins to decrease rapidly.

To determine the transient reverse current in the final stage of the recovery process, known as the reverse phase, we can obtain the value of the hole density gradient near a p-n junction, without solving in full the diffusion equation but using the "compatibility condition" (1.48) of the initial and boundary conditions in the two-dimensional Laplace-Carson transformation. Then, the initial distribution is of the form given by Eq. (2.86) except that $\mathcal{J} = \mathcal{J}_1'$, and the boundary condition is $p(0,\mathcal{J}) = 0$ when $\mathcal{J} \geqslant \mathcal{J}_1'$.

The general solution can be obtained only in the integral form [68] and we shall not give it here because it is cumbersome. Figure 2.33 gives the results of a numerical calculation of the dependence of the transient reverse current for several values of the voltage front rise time. The same figure includes the experimental oscillograms obtained for planar germanium diodes. The considerable difference between the theoretical and experimental amplitudes of the reverse current at time $\mathcal{J}_1'$, equal to or somewhat smaller than b, is due to the departure of the switching pulse front from linearity. The fronts of real pulses are more accurately represented by an exponential function, which differs considerably from the linear function at high values of the argument. The theoretical and experimental curves representing the decay of the reverse current are in satisfactory agreement. When the value of b is increased, the theoretical curves practically merge with the experimental dependences (within the limits of the experimental error in oscillographic measurements).

It is worth noting that in the tail part of the transient characteristic $(\mathcal{J} \gg b)$ all the curves in Fig. 2.33 converge. Thus, when we investigate time intervals far from the beginning of the switching pulse, we no longer need to use reverse pulses with short front rise times.

Let us now return to the first phase of a transient process and consider the dependence of the duration of the recovery phase on the steepness of the front of the reverse pulse. When the front rise time, represented by the sum $a + b$, is increased, the recovery phase becomes shorter (curve 1 in Fig. 2.33) and it decreases to a point when $b = \mathcal{J}_1$. Assuming that $a = b$ and that $(a + b) = \mathcal{J}_{\mathrm{pf}}$, where $\mathcal{J}_{\mathrm{pf}}$ is the pulse front rise time, we obtain from Eq. (2.87) an equation for the determination of the value of the minimum front rise time $\mathcal{J}_{\mathrm{pf}}'$ of the switching pulse at which the recovery phase vanishes:

$$\frac{2\mathcal{J}'_{pf}-1}{\mathcal{J}'_{pf}}\operatorname{erf}\sqrt{\mathcal{J}'_{pf}}+\frac{2}{\sqrt{\pi\mathcal{J}'_{pf}}}e^{-\mathcal{J}'_{pf}}+$$

$$+(B-1)\left[\frac{\mathcal{J}'_{pf}-1}{\mathcal{J}'_{pf}}\operatorname{erf}\sqrt{\frac{\mathcal{J}'_{pf}}{2}}+\frac{\sqrt{2}}{\sqrt{\pi\mathcal{J}'_{pf}}}e^{-\frac{\mathcal{J}'_{pf}}{2}}\right]=1. \quad (2.88)$$

For front durations shorter than $\mathcal{J}'_{pf}$ the value of the peak reverse current does not increase but a narrower or wider region with a constant reverse current is observed in the characteristic. When the front rise time is increased in the range $\mathcal{J}_{pf}>\mathcal{J}'_{pf}$ the value of the peak reverse current decreases monotonically.

When the value of B is sufficiently large and $\mathcal{J}'_{pf}$ is short (<0.5), we can use Eq. (1.57) to expand the error and exponential functions in Eq. (2.87) into series and to obtain $\mathcal{J}'_{pf}$ in an explicit form:

$$\mathcal{J}'_{pf}\simeq\frac{2\pi}{(B+2)^2}. \quad (2.89)$$

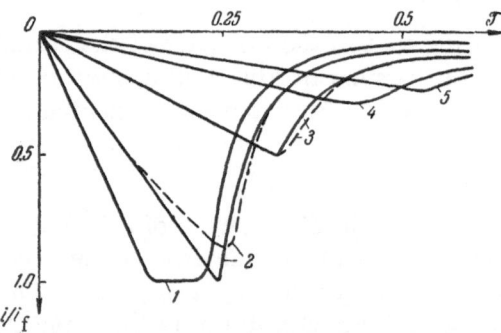

Fig. 2.33. Transient reverse current characteristics for various rise times of the switching pulse front. These characteristics were obtained under the following conditions: B = 1; a = b; 1) b = 0.15; 2) b = 0.275; 3) b = 0.75; 4) b = 0.85; 5) b = 1.3. The dashed curves are the experimental oscillograms.

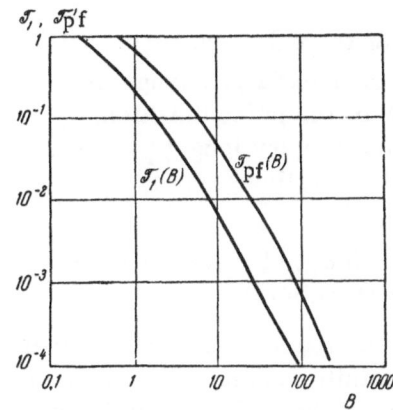

Fig. 2.34. Dependence of the normalized time constants $\mathcal{T}'_{pf}$ and $\mathcal{T}_1$ on the switching conditions (parameter B).

This expression is shown graphically in Fig. 2.34, where, for the sake of comparison, we have included also the dependence of the recovery phase duration $\mathcal{T}_1$ on the parameter B, obtained on the basis of Eq. (2.88) for the instantaneous switching case. We can see that for all values of B > 1, the value $\mathcal{T}'_{pf}$ is several times larger than the value of $\mathcal{T}_1$. Thus, when the exact value of $\mathcal{T}_1$ cannot be measured, because of the finite rise time of the switching pulse, one should reduce the steepness of the pulse front (retaining the linear nature of the rise) until the recovery phase converges to a point. Having thus determined experimentally the value of $\mathcal{T}'_{pf}$, we can use the curves in Fig. 2.34 to find $\mathcal{T}_1$ corresponding to the same value of B. In other words, we can determine indirectly the value of $\mathcal{T}_1$ using pulses with gently rising fronts. For high values of B ($10 < B < 10^2$), we can use the approximate expression for $\mathcal{T}_1$, given in [40]:

$$\mathcal{T}_1 \simeq \frac{0.8}{B^2} \tag{2.90}$$

and Eq. (2.89), to obtain

$$\mathcal{T}'_{pf} \simeq 8\mathcal{T}_1, \quad 10 < B < 10^2. \tag{2.91}$$

If the duration of the forward current pulse cannot be regarded as infinite, the process of accumulation of an excess charge in the base is found to depend on the nature of the front of this pulse. Calculations for the case of a linearly rising front [69], similar to those just given, show that the shape of the pulse front can be ignored if the duration of the flat part of the pulse exceeds $2\tau_p$. A short forward current pulse can be regarded as rectangular if the front rise time does not exceed 20% of the pulse duration. However, if the front rise time is relatively very long ($\mathcal{J}_{pf} > 3$) , then the accumulation of charge in the base follows, in practice, the rise of the current.

A similar analysis has been carried out also for forward current pulses with exponentially rising fronts [70].

Estimates of the distortion of the transient process in a diode due to the finite switching time and the finite resolution time of an oscillograph are given in [67]. It is shown there that when $\mathcal{J}_{pf} < 1$ the various time parameters describing transient processes (with the exception of the initial phase) do not increase by more than $0.25\ \mathcal{J}_{pf}$. This is supported by the curves given in Fig. 2.33, which are plotted on the basis of an exact calculation reported in [68].

Barsukov [71, 72] has calculated the transient process of the postinjection emf decay at high injection levels, and has shown that the error in the determination of the value of u(0) by subtraction of the voltage drop across the base u_b from the total voltage drop (cf. Fig. 2.22) decreases when the decay time of the current pulse is reduced compared with the hole lifetime. However, Barsukov does not give formulas for the calculation of the errors.

§9. MAIN EXPERIMENTAL RESULTS

9.1. Time Dependence of the Transient

Reverse Current

The first experiments aimed at checking the principal conclusions of the theory of transient switching processes in a semiconductor diode were carried out in the middle fifties.

Shulman and McMahon [29] investigated the decay of the transient reverse current through a diode switched in the circuit shown in Fig. 2.24. The investigated samples, which were prepared by growing germanium from the melt, had the following electrical properties in the p- and n-type regions: $\rho_p \approx 0.06-0.07\ \Omega \cdot cm$; $\rho_n = 0.9\ \Omega \cdot cm$; $\tau_p = 20-34\ \mu sec$. The carrier lifetime in germanium was

deduced from the photoconductivity decay. The thickness of the
n-type region in all the diodes was several millimeters and was at
least an order of magnitude greater than the diffusion length of the
holes. The junction area was within the limits 0.2-0.5 mm^2.

When these diodes were switched from a steady-state forward current $i_f = 10$ mA by a reverse voltage step of 10-30 V, very
strong transient reverse currents were observed, which sometimes
reached several amperes. Thus, Shulman and McMahon investigated experimentally the range of the ratio $i(t)/i_f > 10^2$. The initial
amplitude of the reverse current and the amplitude of the switching
pulse were used to determine the total resistance of the circuit at
the beginning of the transient process R_t. It was found that, in all
cases, this resistance was exactly equal to the sum of the load resistance ($R_l = 10$ Ω) and the forward resistance of the base R_b.
The value of R_b was determined experimentally from the slope of
the linear part of the static current-voltage characteristics using
large currents. Moreover, the value of R_b was calculated theoretically taking into account the modulation of the base conductance
by the forward current. The agreement between R_t and ($R_b + R_l$) and
the observation of reverse currents which were hundreds of times
higher than the forward current were of basic importance because
they confirmed the correctness of the theoretically estimated values
of the reverse current peaks (cf. §4). To check its validity, the
analytic dependence $i(t)$, given by Eq. (2.4) in the case of switching
without a limiting resistance, was determined experimentally by

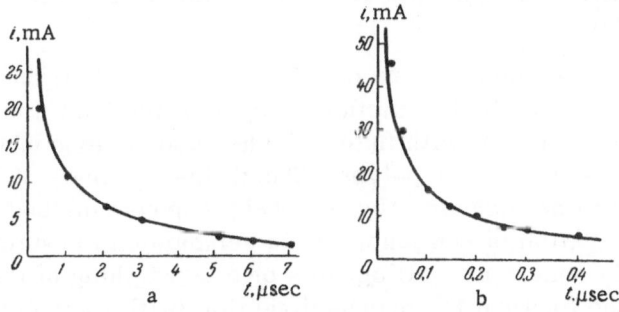

Fig. 2.35. Decay of the reverse current after the switching
of diodes with hole lifetimes $\tau_p = 20$ μsec (a) and 1 μsec
(b) (the load resistance in the circuit was close to zero).

observing it on the screen of an oscillograph. Figure 2.35 gives the
points representing the measurements carried out on diodes in
which the hole lifetimes were 20 and 1 μsec, respectively (a sample with a short hole lifetime was selected from commercial diodes).
The same figure shows also the theoretical curves calculated using Eq. (2.4) for $\tau_p = 20$ μsec and $\tau_p = 0.9$ μsec. We can see
that in both cases the values of τ_p determined from the decay
of the transient reverse current and from the photoconductivity
decay are in excellent agreement. Moreover, the agreement
between the experimental points and the theoretical curves in
a wide range (more than an order of magnitude) of the reverse
current indicates that the functional dependence i(t) given by
Eq. (2.4) is correct.

In plotting the calculated curves, it was assumed that the values of the impressed hole density during the flow of the forward
current were $1.3 \cdot 10^{16}$ cm^{-3} and $0.8 \cdot 10^{16}$ cm^{-3} for the two investigated diodes. Bearing in mind that the minority carrier density in
the base was $n_{n_0} \approx 2 \cdot 10^{15}$ cm^{-3}, it was concluded that the experimental conditions corresponded to a high injection level. The agree-
ment between the experimental and calculated curves under high
injection level conditions was due to the fact that the value of τ_p
was evidently constant throughout the range of investigated injection levels. The influence of the driving field was reduced by Shulman and McMahon by selecting the value of the impressed hole density p_1 in such a way as to obtain the best agreement between the
theory and experiment [in fact, the value of p_1 should have been determined from Eq. (1.17) or from Eq. (125)]. The influence of the high injection level on the experimental reuslts was not discussed by Shulman and McMahon [29].

Similar experiments were carried out by Pell [11] on germanium diodes prepared by two methods: by growing from the melt
and by alloying a crystal with indium. The base regions of the investigated diodes were n or p-type. The diffusion length of the minority carriers was measured by the photoresponse method [73]
after the preparation of p-n junctions in germanium crystals. The
experimentally observed oscillograms of the switching of these
diodes from the forward to reverse direction (with $R_l \rightarrow 0$) were
compared with a theoretical curve of the type given by Eq. (2.4),
and the best agreement between this curve and the experimental
results was used to determine the minority carrier lifetime whose

value was found to be within the limits 130–730 μsec. The diffusion
length, calculated from the transient process, agreed to within
10–20% with the diffusion length found by the photoeffect method;
this was true of all the investigated diodes.

9.2. Investigation of the Recovery Phase

An experimental check of Eq. (2.28), which is the basic equation for the switching of a diode in a circuit with a limiting resistance, has been carried out on numerous occasions in investigations
of various recombination processes in semiconductors.

The first special investigation of the recovery phase was
carried out by Henderson and Tillman [30]. Typical theoretical
and experimental curves for one commercial planar diode are shown
in Fig. 2.36. The theoretical dependence was plotted using Eq.
(2.28), and assuming that τ_p = 8.3 μsec (this value of the hole lifetime in the base was obtain by analysis of the frequency dependence
of the diode impedance under zero bias). The experimental data
can be seen to agree very well with the theoretical curve.

Similar but more extensive and thorough investigations of the
duration of the first switching stage were carried out by Barsukov

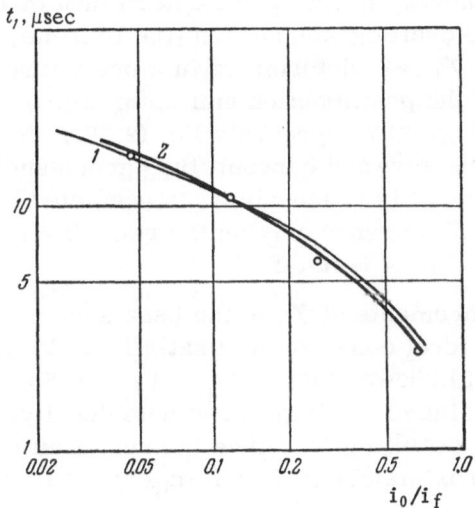

Fig. 2.36. Dependence of t_1 on i_0/i_f: 1) theoretical curve; 2) experimental curve.

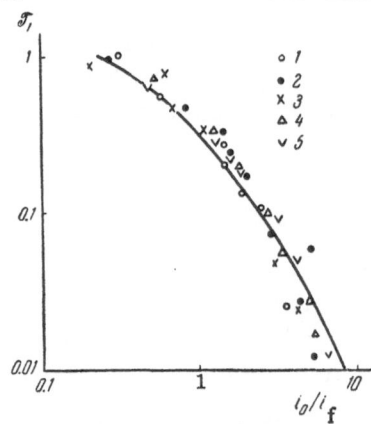

Fig. 2.37. Dependence of the duration of the recovery phase $\mathcal{T}_1$ of a switching transient in an alloyed germanium diode on the ratio i_0/i_f for various values of i_f (mA): 1) 46; 2) 36; 3) 26; 4) 16.5; 5) 7.3. The continuous line is a theoretical curve calculated on the basis of Eq. (2.28).

[48, 51]. In his experiments, the dependence of the recovery phase duration on the ratio $B = i_0/i_f$ was measured for values of B ranging from 0.01 to 10 and for i_f from several milliamperes to tens of amperes. Barsukov used commercial alloyed diode rectifiers of the DGTs23-DGTS27 type* with p-n junction areas of about 2 mm^2 and base thicknesses W close to 0.3 mm. Since the hole lifetimes in all these diodes were not longer than $5\,\mu$sec (which corresponded to the value $L_p = 0.15$ mm), the ratio W/L_p exceeded 2; therefore, in the first approximation, the base could be regarded as semi-infinite (cf. Chap. III).

Figure 2.37 shows the experimental points $\mathcal{T}_1 = f(B)$, obtained for one of the samples using various forward currents [48]. The duration of the recovery phase t_1 was determined directly using time marks on the oscillograph screen [the lifetime, necessary in the calculation of $\mathcal{T}_1$, was determined in a preliminary experiment from the decay of the postinjection emf using Eq. (2.52)]. The continuous curve in Fig. 2.37 represents Eq. (2.28). For all the investigated values of the forward current, the agreement between the theory and experiment is reasonable; as predicted by Eq. (2.28), the time constant $\mathcal{T}_1$ depends only on the ratio $B = i_0/i_f$ and is independent of the value of i_f itself.

In all measurements of $\mathcal{T}_1$ in the base near the p-n junction, the high injection level condition was satisfied. An estimate by means of Eq. (1.25) showed that for $\tau_p \approx 1\,\mu$sec, $S = 2$ mm^2, and $i_f = 7.3$ mA, the value of the impressed hole density near the p-n junction was $p_1 \approx 3 \cdot 10^{14}$ cm^{-3}. Since the resistivity of the base material was approximately 10 $\Omega \cdot$cm ($n_{n0} \approx 2 \cdot 10^{14}$ cm^{-3}), even at

---

* The geometrical dimensions and the electrical properties of p-n junctions and bases of these diodes were similar to those of the D7-type devices.

the minimum values of i_f used in these experiments, the injection level was $\Delta = p_1/n_{n0} \approx 1.5$, and at high values of the forward current the injection level reached $\Delta = 9.5$. These measurements confirmed the correctness of the theoretical calculations of Iglitsyn, Kontsevoi, and their colleagues (cf. § 5), who demonstrated that Eq. (2.28) was also valid at high injection levels provided τ_p was replaced with the hole lifetime under high injection level conditions (τ_∞).

High values of i_f should be applied in the form of forward current pulses in order to avoid overheating a diode. The duration of such pulses should be sufficient to establish a steady-state distribution of the excess density of carriers. On the other hand, the duration of these pulses should be sufficiently short to avoid appreciable heating of the diode in the case of large currents. Control experiments, carried out using forward pulse durations of $40\,\mu$sec (usually the forward pulses were of 10 μsec duration and 50 cps repetition frequency), gave results which were identical with those obtained with 10 μsec pulses.

The dependence of the duration of the first (recovery) phase in the switching process on the ratio of the reverse and forward currents is shown in Fig. 2.38. At each value of the forward current, the dependence is similar to that observed earlier at lower forward currents (Fig. 2.37) but the recovery phase duration t_1 depends also on the forward current itself. Figure 2.38 shows the experimentally determined dependence of the relative change in t_1 on the forward current. Each of the points in that figure was plotted using the data for three diodes. It is evident from Fig. 2.38 that the experimental points are close to the calculated curve and that t_1 varies approximately proportionally to $i_f^{-0.6}$ (for $i_0/i_f = $ const).

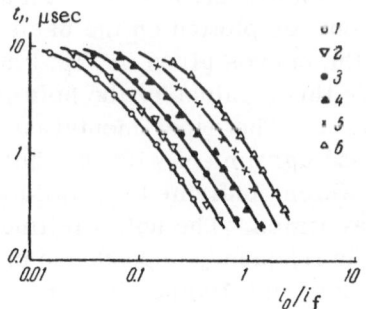

Fig. 2.38. Dependence of the duration of the recovery phase of a switching transient for an alloyed germanium diode on the ratio i_0/i_f for various values of i_f (A): 1) 10.4; 2) 5.4; 3) 2.5; 4) 0.91; 5) 0.28; 6) 0.069.

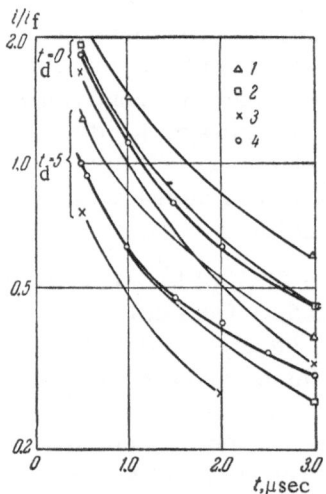

Fig. 2.39. Decay of the transient reverse current for delay times t_d = 0 and 5 μsec. 1,2, and 3 represent theoretical points obtained on the assumption that τ_p = 16.7, 12.5, and 10.0 μsec, respectively; 4) experimental points.

The impressed hole density near the p–n junction may reach values of the order of 10^{17}–10^{18} cm^{-3} for a forward current $i_f \approx 10$ A, which is close to the value of p_{p0} in the recrystallized p–type regions of strutures prepared by the alloying of indium to germanium. This confirms the conclusions given in [51] that the main cause of the improvement in the diode response at high forward densities is the reduction of the injection efficiency of the p–n junction.

9.3. Switching Experiments
with a Delay

Some experimental investigations have been carried out to check the validity of calculations of transient response for the case of switching after a time lag between the end of the forward current pulse and the beginning of the reverse voltage step. Henderson and Tillman [30] investigated specially prepared alloyed diodes with a p–n junction diameter of about 0.3 cm. Since the diffusion length of holes in all their diodes did not exceed 0.025 cm, these samples satisfied completely the requirements of the planar p–n junction. Using a sensitive stroboscope, Henderson and Tillman obtained the transient reverse current decay curves for diodes switched instantaneously and after a time lag t_d = 5 μsec. The time dependences obtained for one of the diodes are given in Fig. 2.39. The same figure includes theoretical curves plotted on the basis of Eq. (2.23), in the same manner as the curves given in Fig. 2.6. The calculated curves were plotted for three values of the hole lifetime in the base: 10, 12.5, and 16.7 μsec. The experimental curves for both switching regimes were in good agreement with the theoretical dependences for τ_p = 12.5 μsec, which confirmed the validity of the dependence i(t) for various delay times. The hole lifetime in the same diode, determined from the frequency dependence of its impedance under zero bias conditions, was 10 μsec, which was

in reasonably satisfactory agreement with the value found by the
pulse method. To check the exact quantitative validity of the formulas, the hole lifetime was determined using three methods: from
the diffucsion length found by the photoresponse method (τ_p = 7.4 to
8.6 μsec): from the photoconductivity decay (τ_p = 8.7μsec); from an
oscillogram of a transient reverse current during switching after
a time lag of 50 μsec (τ_p = 10.5 μsec). The reasonably satisfactory
agreement between these three values of the hole lifetime confirmed
the validity of the theoretical formulas (2.23) and (2.24), including
the coefficients which depend on $\mathcal{J}_d$.

A series of experiments, designed to check the basic relationships governing the transient processes in the case of switching after a time lag, was also carried out by the present author and Postnikova [74].

We used a generator of current pulses whose amplitude could
be varied from 0.1 to 2 A and a generator of voltage pulses of 10-80
amplitude. The decay time of the current pulse did not exceed
0.05μsec and the rise time of the voltage pulse did not exceed 0.03μsec.
The time lag between the pulses could be varied continuously
between the limits of 0: and 8 μsec. A DÉSO-1 wide-band oscillograph
was used to observe the transient response.

This investigation was carried out using alloyed silicon diodes
in which p-n junctions were prepared by alloying an aluminum wire
of 0.3 mm diameter to n-type silicon crystals of 2-7 Ω·cm resistivity. Heat treatment reduced the hole lifetime τ_p in silicon and the
actual degree of reduction could not be predicted. Since the theoretical conclusions could not be checked without the knowledge of the
value of τ_p in the diode base, we used silicon specially doped with
gold in which τ_p was fairly short and practically unaffected by the
heating of silicon during the preparation of a p-n junction. The
hole lifetime τ_p was calculated from the formula

$$\tau_p = \tau_\infty = 1.27 \cdot 10^8 N_{\mathrm{Au}}^{-1} \text{ [sec]},\qquad (2.92)$$

which was derived on the assumption that only the gold acceptor
level took part in the recombination, and the electron and hole capture cross sections were taken from Bemski's paper [162]. As demonstrated in Chap. VII, this assumption is basically wrong: at
high injection levels both gold levels (donor and acceptor) are

active in silicon. However, the quantitative difference between Eq. (2.92) and the exact expression (7.34) is small (less than a factor of 3).

The concentration of gold atoms N_{Au} was determined by radioactivation analysis. The uniformity of the distribution of gold in silicon was checked by the same radioactivation method; the investigation was restricted to crystals in which the inhomogeneity of the distribution of N_{Au} was not greater than 20%. To check the validity of the functional dependence given by Eq. (2.26), we measured the recovered charge Q_{rec} for different delay times t_d when the diode was switched from $i_f = 500$ mA to $U_r = 30$ V. Characteristics of several typical diodes, shown in Fig. 2.40, obeyed exactly the exponential law. To check Eq. (2.26) quantitatively, we measured the recovered charge for batches of diodes with different values of N_{Au} using $i_f = 500$ mA, $U_r = 30$ V, and $t_d = 0.3$ μsec. The results, taking into account the scatter of the values of Q_{rec} for individual diodes, are given in Fig. 2.41, which also includes a calculated curve. Each of the experimental points shown in Fig. 2.41 was the average for a batch of diodes prepared from at least three silicon ingots with the same value of N_{Au}. It is evident from

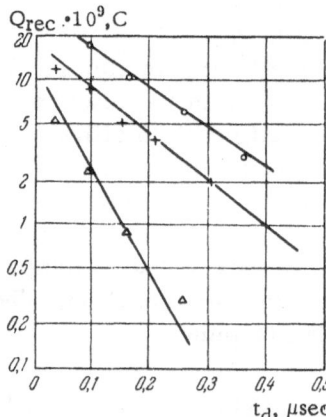

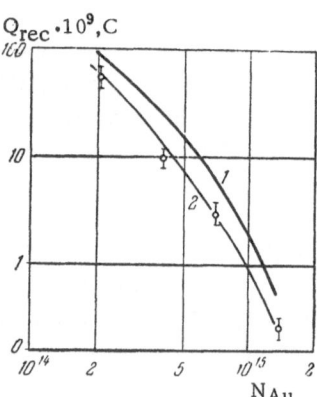

Fig. 2.40. Dependence of the recovered charge on the delay time for $i_f = 500$ mA and $U_r = 30$ V (the three lines represent different diodes).

Fig. 2.41. Dependence of the recovered charge on the concentration of gold in silicon: 1) calculated curve; 2) experimental curve.

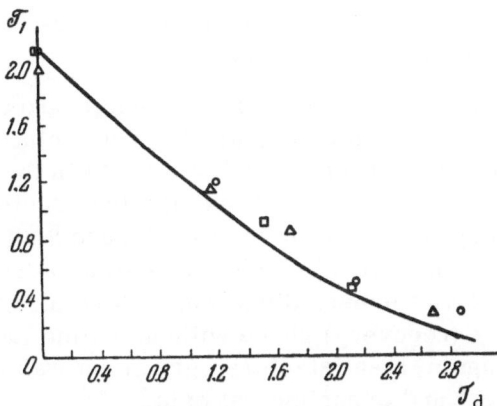

Fig. 2.42. Dependence of the duration of the first
(recovery) phase $\mathcal{J}_1$ on the normalized time $\mathcal{J}_d$.
The continuous curve was calculated using Eq.
(2.40). The experimental points were obtained
using three diodes.

Fig. 2.41 that the experimental curve agreed with the theoretical
dependence to a factor of 2.

Bearing in mind the high injection level conditions in our experiments and that the expression (2.26) was obtained for $\Delta \ll 1$,
and also taking into account the inaccuracy of Eq. (2.92), we concluded that such agreement between the theoretical and experimental dependences was satisfactory. Obviously, the presence of a driving field in the base when $\Delta > 1$ would pull holes away from the p–n
junction and reduce the recovered charge.

We carried out several control experiments. We found that
the recovered charge was a linear function of the forward current
(within the range 100–1500 mA) for all the investigated diodes; this
indicated that the hole lifetime was constant, as assumed in the
theoretical calculations. The dependence of the recovered charge
on the amplitude of the reverse voltage pulse was in the form of a
monotonically rising curve (at low values of U_r) with a saturation
plateau (at high values of U_r). For all the investigated diodes, the
value of $U_r = 30$ V was in the plateau (saturation) region; further
increase of the reverse pulse amplitude did not increase the recovered charge. This indicated that the contribution of the current
during the first phase of the transient to the total recovered charge

was small, i,e., that the boundary condition (1.22), assumed in the derivation of Eq. (2.26), was correct.

Formula (2.40) was checked using diodes prepared without preliminary doping of silicon with gold. We determined the duration of the recovery phase t_1 on the delay time t_d in the case of switching from $i_f = 1$ A to $U_r = 20$ V, using a load resistance $R_l = 500$ Ω, which corresponded to $i_0/i_f = 0.04$. Figure 2.42 shows the experimental points for several diodes, as well as a calculated curve. The value of τ_p for each diode was determined from the duration of the first (recovery) phase without a time lag, using Eq. (2.28). Once again, a satisfactory agreement was obtained between the theoretical and experimental data.

9.4. Observation of Postinjection EMF Decay

The first experimental investigation of the postinjection emf decay was carried out by Gossick [52].

Gossick studied germanium planar diodes as well as the emitter and collector p-n junction of alloyed germanium transistors. For all the investigated samples (12 diodes and 9 transistors), he found the theoretically predicted linear decay of u(t) as long as the inequality u(t) > kT/q was satisfied. The slopes of the linear part were practically indentical for the emitter and collector junctions of the transistors. The linearity of the transient response was retained when the forward current pulse amplitude was changed by a factor of 10^4. The postinjection emf decay was investigated for two diodes at very low values of the emf [u(t) $\ll$ kT/q]. It was found that the decay u(t) in the part of the characteristic was described satisfactorily by an exponential dependence on time. It was remarkable that the values of the hole lifetime τ_p, calculated from the linear and exponential parts of the decay curves using Eqs. (2.52) and (2.55), agreed to within 10%.

Lederhandler and Giacoletto [54] compared the lifetimes of holes, calculated from the decay of the postinjection emf of a p-n junction (the linear region), with the values obtained from the photoconductivity decay in germanium crystals illuminated with light pulses [75]. The agreement was satisfactory in all cases, except when the values of τ_p were very large ($\sim$700 μsec). The disagreement in the case of long lifetimes could be due to a strong influence of the surface recombination on the effective lifetime in the base.

Lederhandler and Giacoletto also investigated a large number of p-n junctions in alloyed transistors during the forward current flow. In contrast to Gossick [52], Lederhandler and Giacoletto found that at high forward currents, corresponding to high injection levels in the base, the dependence of the voltage across the p-n junction consisted of two linear parts with different slopes. The slope of the linear region at high values of t was equal to the slope of the linear region observed in the case of small forward currents and low injection levels. Thus, even in the case of large forward current pulses, it was possible to find a part of the postinjection emf decay curve suitable for the determination of the value of τ_p at low injection levels.

In practice, however, it was more convenient to select such a forward current that the dependence observed on the oscillograph screen was a straight line with a single slope.

Curtis and Gossick [76] investigated the postinjection voltage decay in surface-barrier germanium structures. The rectifying contact was prepared by evaporation of a gold film on a carefully etched surface of an n-type germanium crystal of about 5 Ω·cm resistivity. In these measurements, the current pulses were 20 μsec long and of 200 mA amplitude. The experimentally determined postinjection voltage decay curves are shown in Fig. 2.43 for two samples. These samples differed in the value of the injection efficiency and the height of the potential barrier in the contact region.

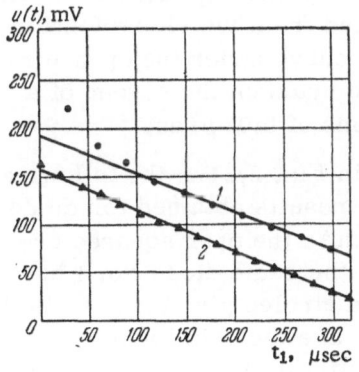

Fig. 2.43. Decay of the postinjection voltage for two surface-barrier diodes with I_f = 200 mA: 1) diode with γ = 0.92; 2) diode with γ = 1.

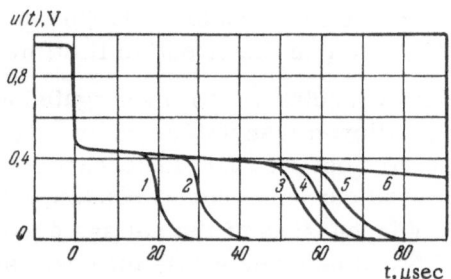

Fig. 2.44. Decay of the postinjection emf in
a diode with a grown p-n junction and τ_p =
9 μsec, if = 10mA, and various load resistance
R_l(kΩ): 1) 0.43: 2) 0.86: 3) 6.8; 4)10; 5) 30;
6) ∞.

Both diodes had rectilinear regions in the u(t) curves, which were
reached sometime after the switching-off of the forward current.
The hole lifetimes, calculated from these regions using Eq. (2.52),
were 64-67 μsec. Measurements carried out on the same germanium crystals using the injection method [73] gave the values τ_p =
65-70 μsec, in excellent agreement with the results of the pulse
method.

The effect of the load resistance in the diode circuit on the
postinjection emf decay was investigated experimentally by Mokeev.
He studied three different silicon planar diodes: No. 1: a diffused
mesa diode with a p-n junction area of 10^{-5} cm^2 and a hole lifetime
of 0.12 μsec; No. 2: a diode with a p-n junction prepared during
growth of an ingot from the melt (S = 4 mm^2 and τ_p = 9 μsec); No. 3:
a diffused mesa diode with S $\approx$ 5 $\cdot$ 10^{-4} cm^2 and τ_p = 2.2 μsec. The
lifetimes in all samples were determined from the slope of the lin-
ear part of the postinjection emf decay curve under the open-circuit conditions. The observations were made on the screen of a
DÉSO-1 oscillograph with scale divisions of 0.01 μsec.

Mokeev's results are presented in Fig. 2.21 in the following
form: the black squares represent the results obtained for diode
No. 1 using R_l = 430 Ω and i_f = 10-100 mA; the open squares are
used to denote the results obtained for the same diode No. 1 but
using i_f = 20 mA and R_l = 430-∞ Ω; the circles give the results obtained for diode No. 2, and the triangles are the data for diode No.3.
The initial value of the emf, u(0), for diffused diodes was within

the limits 0.7–0.8 V, and for diodes grown from the melt, 0.45 V.
Thus, the experimental values of u(0) were close to those which
were used in plotting the calculated curves in Fig. 2.21. The satisfactory agreement between the experimental and theoretical data
confirmed the validity of Eqs. (2.60) and (2.62); the latter equation
follows from the former and is used to determine the value of $R_{l\infty}$.

Figure 2.44 shows oscillograms of the transient characteristics of diode No. 2 for various load resistances. The time constant
of the second (exponential) phase of the transient increased by a
factor of 3 when the load resistance was increased by a factor of
70. In the first approximation, we could assume that Eq. (2.61)
was justified experimentally; some disagreement was evidently
due to the inaccuracy of the model used in the calculations.

Chapter III

Planar Diode with a Thin Base

In many types of planar semiconductor diode, the assumption of an infinitely thick base is not satisfied. A parallel ohmic contact with the n-type region is usually located in the immediate vicinity of a p-n junction and the base resistance is reduced by the shortening of the distance between this contact and the rectifying p-n junction. In those cases when the distance between the p-n junction and the ohmic contact is comparable with the hole diffusion length, the presence of such a contact alters the process of accumulation and dispersal of the excess charge and, consequently, the nature of the transient processes which accompany the switching of such a diode.

§ 10. STEADY-STATE DISTRIBUTION OF
HOLES IN THE BASE

We shall consider the model of a diode shown in Fig. 3.1. We shall assume that the planar rectifying and ohmic contacts are infinite, plane-parallel, and separated by a distance W, i.e., we shall reduce this problem to the one-dimensional case.

We shall assume that the p-n junction plane coincides with the plane x = 0 and that the positive direction of the x axis is directed into the base so that the ohmic contact lies at x = W. As before, the n-type region will be called the base; in this case, W represents the base thickness. We shall retain all the assumptions about p-n junctions made in the preceding chapters and, thus, the boundary conditions at the p-n junction will be assumed to obey, as before, Eqs. (1.20) and (1.21), i.e., when we know the current flowing through the p-n junction, we know the hole density gradient at

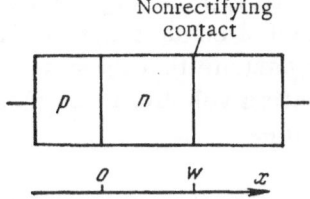

Fig. 3.1. Model of a thin-base
diode

x = 0 (a boundary condition of the second kind) and when a definite
voltage is applied to the p–n junction, we also know the value of the
hole density at x = 0 (a boundary condition of the first kind).

10.1. Types of Nonrectifying Contact

To analyze mathematically the transient processes in a thin-
base diode it is necessary to determine the boundary conditions in
the ohmic contact plane.

There is as yet no agreed theory of the nature of nonrectifying contacts in semiconductor devices. Ideal models of real nonrectifying contacts are of two types: semiconductor–metal and
n–n$^+$ contacts, where the latter represents a boundary between two
regions with very different majority carrier densities within a semiconductor crystal of one conduction type (n-type).

A semiconductor–metal boundary is characterized by a surface recombination velocity of holes S_R which is introduced by the
following relationship:

$$-D_p\left(\frac{\partial p}{\partial x}\right)_W = S_R(p_W - p_{n0}),\qquad(3.1)$$

where the subscript W means that the variable is measured at
x = W. Thus, at the semiconductor–metal boundary the diffusion
current of holes is proportional to the surface recombination velocity and the hole density in the plane of this boundary. Ohmic
semiconductor–metal contacts described by Eq. (3.1) are known as
the recombination–type contacts.

Often, in a wide range of currents flowing through a recombination–type contact, the hole density near such a contact is close

to the equilibrium value, i.e., $p_W \approx p_{n0}$. In this case, to obtain a finite current through such a contact and to satisfy the condition (3.1), it is necessary to assume that $S_R = \infty$. In the case of an infinite surface recombination velocity at an ohmic contact, the boundary condition becomes

$$p_W - p_{n0} = 0 \quad \text{when} \quad x = W. \tag{3.2}$$

The current is transported across an n-n$^+$ junction mainly by electrons, i.e., by the majority carriers; thus, n-n$^+$ junctions are noninjecting contacts. However, since some holes do escape from the base through such a contact, the boundary condition in the x = W plane then becomes:

$$j_p = qvp \quad \text{when} \quad x = W, \tag{3.3}$$

where v represents the rate of loss of the minority carriers through the n-n$^+$ junction. The condition (3.3) differs from (3.1) in that (3.3) contains the total hole current density, given by the general equation (1.3) and not only the diffusion component, as in Eq. (3.1). At low injection levels, the rate of hole leakage is constant [77, 78]:

$$v = \sqrt{\frac{D_p^+}{\tau_p^+} \cdot \frac{n_{n0}}{n_{n0}^+}}, \tag{3.4}$$

where the superscript "+" indicates the n$^+$ region. We can easily see that when $n_{n0}^+ \gg n_{n0}$, it follows from Eq. (34) that $v \rightarrow 0$ and, therefore, the boundary condition for an ideal noninjecting contact is

$$j_p = 0. \tag{3.5}$$

Comparison of Eqs. (3.2) and (3.5) shows directly the basic difference between the influence of different contacts on transient processes in a thin-base diode.

A recombination-type ohmic contact is a perfect absorber of holes near the p-n junction and this naturally reduces the accumulation of holes in the base and, consequently, accelerates the transient processes.

A noninjecting contact represents a perfect reflector of holes and, therefore, when such a contact lies near a rectifying p-n junction, all the carriers injected by the junction are concentrated in the direct vicinity of the junction, but when $W \to \infty$, excess carriers are distributed throughout the semi-infinte base. An increase in the number of holes near the p-n junction naturally increases the switching inertia of the diode (increases its response time).

10.2. Ohmic Recombination-Type Contacts

Knowing the boundary conditions at $x = 0$ and $x = W$, we can easily find the steady-state distributions of excess carriers in the base during the flow of a forward current of density j_f. The general solution for a thin-base diode and a recombination-type ohmic contact has been obtained by Penin[79]. He has used the duffusion equation (1.13) for a low injection level and a low base resistance (i.e., he has assumed that there is no electric field in the base), as well as the boundary conditions (1.21) at $x = 0$ and (3.1) at $x = W$. The steady-state hole distribution in the base during prolonged flow of the forward current is of the form (this can be checked easily by direct substitution):

$$p(x) = p_{n0} + (p_1 - p_{n0}) \frac{D_p \cosh \dfrac{x-W}{L_p} - S_R L_p \sinh \dfrac{x-W}{L_p}}{D_p \cosh \dfrac{W}{L_p} + S_R L_p \sinh \dfrac{W}{L_p}} \qquad (3.6)$$

where p_1 is related to the voltage across the p-n junction by Eq. (1.21). The density of the forward current through the p-n junction is obtained from Eq. (1.20) by differentiating Eq. (3.6) with respect to the coordinate at $x = 0$; this current density is

$$j_f = \frac{qD_p(p_1 - p_{n0})}{L_p} \frac{D_p \sinh \dfrac{W}{L_p} + S_R L_p \cosh \dfrac{W}{L_p}}{D_p \cosh \dfrac{W}{L_p} + S_R L_p \sinh \dfrac{W}{L_p}}. \qquad (3.7)$$

Thus, the simple expression (1.25), relating j_f with p_1 for a thick-base diode, does not apply to a thin-base diode although the direct proportionality of j_f and p_1 is still retained (when the value of p_{n0} is ignored as being small compared with p_1).

For an infinite recombination velocity in the plane of an ohmic contact, the expressions (3.6) and (3.7) simplify to:

$$p(x) = p_{n0} + (p_1 - p_{n0}) \frac{\sinh \frac{W - x}{L_p}}{\sinh \frac{W}{L_p}} \simeq p_1 \frac{W - x}{W}, \qquad (3.8)$$

$$j_f = \frac{qD_p(p_1 - p_{n0})}{L_p} \coth \frac{W}{L_p} \simeq \frac{qD_p p_1}{W}. \qquad (3.9)$$

The approximate equalities in Eqs. (3.8) and (3.9) are obtained on the assumption that $W/L_p \ll 1$ and $p_{n0} = 0$. Thus, in a diode with a very thin base* and an ideal recombination-type ohmic contact, the hole density decreases linearly from the value p_1 at the p-n junction to zero at the contact. The current density in a diode with $W/L_p \ll 1$ is L_p/W times larger than in a "long" diode for the same voltage drop across the p-n junction. When the current densities are equal, the impressed density of holes at the junction in a thin-base diode is L_p/W times lower than in a thick-base diode.

The steady-state hole distribution in a thin-base diode has been obtained by Stafeev [16] for a high injection level by solving Eq. (1.18) with the boundary conditions given by Eqs. (1.21) and (3.2). The identity of Eqs. (1.18) and (1.13) means that, in the case $\Delta \gg 1$, the hole distribution in the base is described by Eq. (3.8) in which L_p is replaced l_p, where l_p is defined in Chap. I, § 2 [immediately after Eq. (1.18)]. It follows from Eq. (1.3) that when $p \gg n_{n0}$, the diffusion and drift components of the current are equal and, therefore, at a high injection level, a thin-base diode obeys

$$j_f \simeq q p_1 \frac{2D_p}{l_p} \coth \frac{W}{l_p}, \qquad (3.10)$$

---

* When we speak of a thick or thin base, we shall not mean the absolute thickness W but the ratio of this thickness and the diffusion length of holes. We shall use the following terminology: a thick-base diode, a diode with a semi-infinite base, or a long diode when W/L_p can have any value or when $W/L_p \gg 1$; a thin-base diode when $W/L_p \approx 1$; a diode with a very thin base when $W/L_p \ll 1$.

which differs from Eq. (3.9) only by a factor of 2 and the replacement of L_p with l_p.

10.3. Ohmic Noninjecting Contacts

The steady-state hole distribution in the base was obtained by Baranov [17, 80] for a low injection level by solving Eq. (1.11) for $\partial p/\partial t = 0$, taking into account the presence of an electric field. Using the boundary conditions (1.21) and (3.3), Baranov obtained

$$p(x)=(p_1-p_{n0}) \exp\left(\frac{\mu_p E \tau_p}{2L_p^2} x\right)[e^{-gx}+2ae^{-gW} \sinh gx], \qquad (3.11)$$

where

$$a=\frac{D_p g + \dfrac{\mu_p E}{2} - v + \mu_p E \ \dfrac{p_{n0}}{p_1-p_{n0}} \exp\left(-\dfrac{\mu_p E \tau_p W}{2L_p^2} + gW\right)}{2D_p g \cosh gW - \mu_p E \sinh gW + 2v \sinh gW}$$

and

$$g=\sqrt{\left(\frac{\mu_p E \tau_p}{2L_p^2}\right)^2 + \frac{1}{L_p^2}}.$$

We can easily see that if the base resistivity is not very high and the influence of the electric field can be neglected (this can be done when $E \ll kT/qL_p$), the distribution given by Eq. (3.11) reduces to that given by Eq. (3.6) provided we replace S_R with v.

When the leakage of holes through an n–n⁺ junction is very small ($S_R = 0$), the expressions (3.6) and (3.7) become

$$p(x)=p_{n0}+(p_1-p_{n0})\frac{\cosh\dfrac{x-W}{L_p}}{\cosh\dfrac{W}{L_p}} \simeq p_1, \qquad (3.12)$$

$$j_f=\frac{qD_p(p_1-p_{n0})}{L_p} \cdot \tanh \frac{W}{L_p} \simeq \frac{qD_p p_1}{L_p} \frac{W}{L_p} = \frac{qp_1 W}{\tau_p}. \qquad (3.13)$$

The right-hand sides of Eqs. (3.12) and (3.13) are valid for diodes with very thin bases on the assumption that $p_{n0} = 0$.

Thus, in diodes with ideal noninjecting contacts and very thin bases ($W/L_p \gg 1$), the hole distribution in the base is uniform and the impressed hole density is p_1 given by Eq. (1.21). The current density in such a diode is L_p/W times smaller than in a similar thick-base diode for the same voltage drop across the p-n junction. When the current densities are equal, the impressed hole density near a p-n junction in a thin-base diode with a noninjecting ohmic contact is L_p/W times higher' than in a long diode.

Thus, in the case of a low injection level and a low base resistance, the steady-state hole distribution in the base of a diode with either of the two types of ohmic contact is given by Eq. (3.6) during the flow of a forward current: for an ideal semiconductor-metal contact, we have $S_R = \infty$ and Eq. (3.6) reduces to Eq. (3.8), while for an ideal $n-n^+$ junction $S_R = 0$ and Eq. (3.6) reduces to Eq. (3.12).

The steady-state distribution of excess holes in a thin-base diode at a high injection level has been deduced by Baranov [17] and it does not differ basically from Eq. (3.6).

10.4. Refinement of Boundary Conditions

In the case of large forward current densities, the assumption of Eq. (3.2) that the hole density near a recombination-type ohmic contact has the equilibrium value may violate the quasineutrality condition which has been used to derive Eq. (1.18), i.e., it may give rise to an internal inconsistency in the theory. In fact, very high current densities give rise, in accordance with Eq. (1.3), to large divergences of the electric field in the contact region because the condition (3.2) implies that the hole current is equal to zero in the x = W plane (or when x > W). However, according to Poisson's equation (1.8), a high value of $\partial E/\partial x$ unavoidably produces a space charge which, under some conditions, may exceed the charge of equilibrium carriers and violate the quasineutrality condition.

To avoid this internal inconsistency, Sokolov [81] has suggested that some hole leakage does take place through an ohmic contact and that the boundary conditions in the x = W plane should be represented in the form

$$j_p(x=W) = qS_R(p_W - p_{n0}) + \frac{j}{K} p_W, \qquad x = W, \qquad (3.14)$$

where K is some parameter depending on the properties of the contact and on the properties of the quasineutral region next to the contact. In each specific case, the coefficient K must be determined experimentally. At very high injection levels, the first term on the right-hand side of Eq. (3.14) can be dropped because it is small.

The current-voltage characteristic of a thin-base diode [81], calculated using Eq. (1.18) and the nonlinear boundary condition (3.14), is similar to the characteristic obtained by Stafeev (16] who used the boundary condition (3.2). The estimates obtained by Baranov [17] show that the use of Eq. (3.14) instead of Eq. (3.12) makes a perceptible difference only when $W/L_p \ll 1$ and $\Delta = p_1/n_{n0} > 10^2$, i.e., under conditions which are rarely realized in practice.

A different nonlinear boundary condition for the ohmic contact plane has been proposed by Tkhorik [82] in order to explain the properties of thin-base diodes. Since the surface recombination velocity in germanium depends on the injection level (this has been proved theoretically and experimentally), we can postulate a similar dependence for the recombination velocity S_R at an ohmic contact. Using the published experimental data for the basic parameters of a ground p-type germanium surface, Tkhorik has plotted the dependence of S_R on Δ for this case. It is evident from Fig. 3.2 that this dependence can be approximated by the following expresion, which is valid in a wide range of injection levels:

$$S_R = \Gamma/\sqrt{\Delta}, \qquad (3.15)$$

where Γ is a constant related to the properties of a semiconductor and the quality of the surface treatment.

Experiments carried out on diodes with bases of various thicknesses and with different ohmic contacts have shown that Tkhorik's relationship applies also to p-p$^+$ junctions in germanium. In this case, Δ is the injection level near a nonrectifying contact.

The use of the boundary condition (3.1), combined with the relationship (3.15), may give more accurate results than the use of Eq. (3.1) with S_R = const.

10.5. Stored Charge

It is very interesting to compare the values of the charge stored in various thin–base diodes since the stored charge governs, to a considerable extent, the response of a diode during switching. The stored charge Q_{st} is found by integrating the excess hole density over the whole volume of the base:

$$Q_{st} = \int\limits_0^W Sq \left[p(x) - p_{n0} \right] dx. \qquad (3.16)$$

Substituting into Eq. (3.16) the general hole density distribution, given by Eq. (3.6), and making self–evident transformations, we obtain

$$Q_{st} = Sq \left(p_1 - p_{n0} \right) L_p \frac{D_p \sinh \dfrac{W}{L_p} + S_R L_p \left(\cosh \dfrac{W}{L_p} - 1 \right)}{D_p \cosh \dfrac{W}{L_p} + S_R L_p \sinh \dfrac{W}{L_p}}. \qquad (3.17)$$

Using Eq. (3.7) and introducing dimensionless coefficients

$$\frac{W}{L_p} = W_n \text{ and } \frac{S_R L_p}{D_p} = \Pi, \qquad (3.18)$$

we can transform Eq. (3.17) to the following form

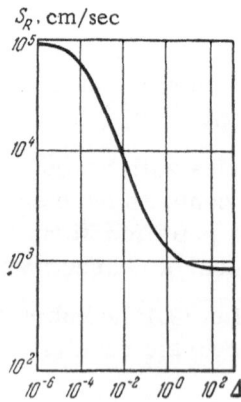

S_R, cm/sec

Fig. 3.2. Dependence of the surface recombination velocity on the injection level for p–type germanium with $\rho = 2.2 \ \Omega \cdot cm$.

$$Q_{st} = i_f \tau_p f(W_n, \Pi), \qquad (3.19)$$

where

$$f(W_n, \Pi) = \frac{\sinh W_n + \Pi(\cosh W_n - 1)}{\sinh W_n + \Pi \cosh W_n} . \qquad (3.20)$$

As in the planar diode case, we can express Q_{st} in terms of the product $i_f \tau_p$; the only difference is the introduction of the coefficient $f(W_n, \Pi)$, whose values lie between 1 and 0 for all possible values of the arguments. Figure 3.3 shows graphically the dependences of the relative value of Q_{st} on the parameter $S_R L_p/D_p$ for various values of the ratio W/L_p. The curve for each of the values of W/L_p has two horizontal parts: at low and high values of Π. It is important to note that at low values of Π (i.e., at low values of the surface recombination velocity), $Q_{st} = i_f \tau_p$ for all diodes and this value is equal to the charge stored in a semi-infinite base. The value of Q_{st} in the second horizontal part depends on the value of W/L_p.

In the limiting case, $S_R \to \infty$, the expression for the charge stored in a diode with a very thin base becomes

$$Q_{st} = i_f \tau_p (1 - \operatorname{sech} W_n) \simeq i_f \tau_p \frac{W_n^2}{2} = \frac{i_f W^2}{2 D_p}, \qquad (3.21)$$

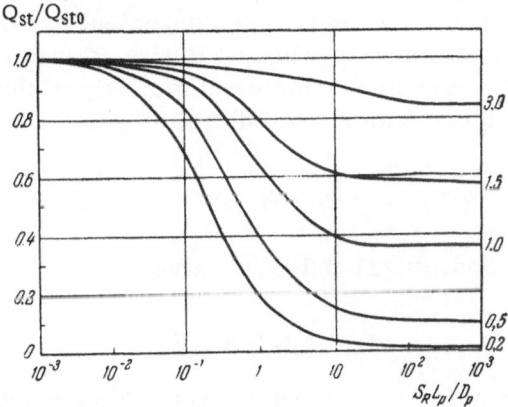

Fig. 3.3. Dependence of the stored charge on the parameter $S_R L_p/D_p$ for diodes with various normalized base thicknesses W/L_p.

p, arbitr. units

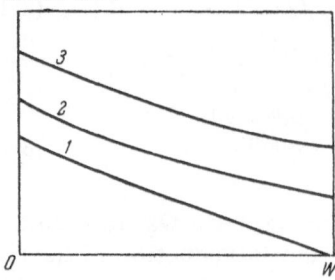

Fig. 3.4. Distribution of the hole density during flow of a steady-state forward current in the base of a diode with $W/L_p = 1$ and different values of S_R: 1) ∞; 2) D_p/L_p; 3) 0.

where the approximate equality is valid when $W_n \ll 1$. It follows from Eq. (3.21) that when the ratio W/L_p is reduced, the value of the stored charge decreases proportionally to $(W/L_p)^2$; moreover, for a diode with a very thin base the stored charge ceases to depend on the hole lifetime and is governed only by the forward current and the base thickness. Physically, this is explained as follows: when $W/L_p \ll 1$ and $S_R \to \infty$, the rate of annihilation of excess holes in the base is governed primarily by the recombination at the ohmic contact and not in the interior of the semiconductor. Naturally, the importance of the contact recombination increases when the ratio W/L_p is reduced. In the other extreme case ($S_R \to 0$), the stored charge is independent of the base thickness and equal to $i_f\tau_p$. Several expressions for the charge stored in a thin-base diode [including Eq. (3.21)] have been obtained by Rediker et al. [31, 83].

To obtain an approximate qualitative estimate of the response of a thin-base diode it is sometimes convenient to use the concept of the effective lifetime τ_{eff}, whose value is defined so that the stored charge is given by a simple relationship of the same type as in the case of a diode with a semi-infinite base:

$$Q_{st} = i_f \, \tau_{eff}. \tag{3.22}$$

Comparison of Eqs. (3.22) and (3.19) gives

$$\tau_{eff} = \tau_p f(W_n, \Pi) \tag{3.23}$$

and, consequently, for a diode with a very thin base and $S_R = \infty$, $\tau_{eff} = W^2/2D_p$, i.e., it is close to the average time taken by holes to diffuse from the p-n junction to the ohmic contact. For diodes in which the surface recombination velocity in the plane of the ohmic contact is low, we have $\tau_{eff} = \tau_p$.

The response of a diode under given switching conditions is governed not only by the charge of excess holes stored in the base but also by the nature of the distribution of this charge. The higher the fraction of the charge stored in the direct vicinity of the p-n junction, the slower are the transient switching processes. It is worth noting that the effective lifetime τ_{eff}, introduced in Eq. (3.22), gives only the value of the stored charge and cannot provide an exact quantitative criterion of switching processes in thin-base diodes. The influence of the surface recombination velocity in the ohmic contact plane on the nature of the hole distribution in the base is illustrated in Fig. 3.4, which gives the dependences p(x) for three diodes with $S_R = 0$, $S_R = D_p/L_p$, and $S_R = \infty$. For all three diodes, we have assumed $W/L_p = 1$ and the same forward current density.

In the case when $S_R = D_p/L_p$, it is evident from Eq. (3.6) that the hole distribution obeys the simple law $p(x) = p_1 \exp(-x/L_p)$, i.e., exactly the same law which would apply to a diode having an ohmic contact at infinity. Under all switching conditions, a diode with $S_R = D_p/L_p$ behaves, irrespective of the base thickness, exactly as a diode with a semi-infinite base.

Thus, this value of S_R is , in a certain sense, a limit: if $S_R > (D_p/L_p)$ the response of a diode improves with decreasing base thickness, but if $S_R < (D_p/L_p)$ the response deteriorates when the distance between the ohmic and rectifying contacts is reduced.

§ 11. SWITCHING WITHOUT A RESISTANCE
IN THE DIODE CIRCUIT

11.1. Method for Solving the Diffusion Equation

We shall find the distribution of holes in a planar diode with an arbitrary base thickness during the switching from a steady-state forward current by a reverse voltage applied instantaneously to the p-n junction, i.e., when the resistance in the diode circuit is $R_l = 0$. We shall consider a diode with an ideal ohmic contact of the recombination type ($S_R = \infty$) because this case differs most from the thick-base conditions. The solution for a diode with a noninjecting ohmic contact can be obtained in a similar manner.

Thus, we shall solve the diffusion equation (1.13) with an initial distribution given by Eq. (3.8) and the zero boundary conditions given by Eqs. (1.22) and (3.2).

Assuming, for the sake of simplicity, that $p_{n0} = 0$ and using, as before, the substitution defined by Eq. (1.32), we obtain the diffusion equation and the boundary conditions in the following form:

$$\frac{\partial \varphi}{\partial \mathcal{J}} = \frac{\partial^2 \varphi}{\partial X^2}, \tag{1.33}$$

$$\left. \begin{array}{ll} \varphi(0, \mathcal{J}) = 0, & \mathcal{J} > 0, \\ \varphi(W_n, \mathcal{J}) = 0, & \mathcal{J} > 0, \\ \varphi(X, 0) = p(X, 0) = p_1 \dfrac{\sinh(W_n - X)}{\sinh W_n}, & \mathcal{J} = 0. \end{array} \right\} \tag{3.24}$$

The general method of integration of this equation in partial derivatives over a finite segment is based on the separation of variables, i.e., the required solution is represented in the form of a product of two functions, one of which depends only on X and the other only on $\mathcal{J}$, i.e.,

$$\varphi(X, \mathcal{J}) = \Phi_1(X) \cdot \Phi_2(\mathcal{J}). \tag{3.25}$$

Substituting the suggested solution (3.25) into Eq. (1.33) and dividing both sides of the equation by $\Phi_1 \Phi_2$, we obtain

$$\frac{1}{\Phi_1} \frac{d^2 \Phi_1}{dX^2} = \frac{1}{\Phi_2} \frac{d\Phi_2}{d\mathcal{J}}. \tag{3.26}$$

This equation should be an identity, i.e., it should apply to all possible values of the independent variables $0 < x < W_n$ and $\mathcal{J} > 0$. Since the right-hand side of the above equation is only a function of the variable $\mathcal{J}$, and the left-hand side is only a function of X, Eq. (3.26) can be an identity only if both sides of the equation remain constant when their arguments vary, i.e.,

$$\frac{1}{\Phi_1} \frac{d^2 \Phi_1}{dX^2} = \frac{1}{\Phi_2} \frac{d\Phi_2}{d\mathcal{J}} = -\lambda, \tag{3.27}$$

where λ is a constant. The minus sign in front of λ is inserted only for convenience in later analysis.

Equation (3.27) for Φ_1 and Φ_2 yields the following ordinary differential equations:

$$\frac{d^2\Phi_1}{dX^2} + \lambda\Phi_1 = 0, \tag{3.28}$$

$$\frac{d\Phi_2}{d\mathcal{J}} + \lambda\Phi_2 = 0 \tag{3.29}$$

with the boundary conditions

$$\Phi_1(0) = \Phi_1(W_n) = 0. \tag{3.30}$$

The particular solution of Eq. (3.28) with the boundary conditions given by Eq. (3.30) is

$$\Phi_1^{(m)} = \sin\sqrt{\lambda_m}X, \tag{3.31}$$

and it is known [19] that this solution is not identically equal to zero only when $\lambda_m = (\pi m/W_n)^2$, where m is any integer. Naturally, we should seek the particular solutions of Eq. (3.29) only for these values of λ_m. A simple substitution shows clearly that the functions

$$\Phi_2^{(m)} = C_m e^{-\lambda_m \mathcal{J}} \tag{3.32}$$

are the solutions of Eq. (3.29) for any values of the constant of integration C_m.

The general solution of Eq. (1.33) is the sum of all the possible particular solutions $\Phi_1^{(m)}\Phi_2^{(m)}$, i.e., it can be represented in the form of an infinite series

$$\varphi(X, \mathcal{J}) = \sum_{m=1}^{\infty} C_m \exp\left[-\left(\frac{\pi m}{W_n}\right)^2 \mathcal{J}\right] \sin\frac{\pi m}{W_n} X. \tag{3.33}$$

The reverse substitution of $\varphi(X, \mathcal{J})$ by $p(X, \mathcal{J})$, on the basis of Eq. (1.32), yields the general solution for our problem:

$$p(X, \mathcal{J}) = \sum_{m=1}^{\infty} C_m \exp\left\{-\left[\left(\frac{\pi m}{W_n}\right)^2 + 1\right]\mathcal{J}\right\} \sin\frac{\pi m}{W_n} X. \tag{3.34}$$

The coefficients C_m are found from the requirement that the solution obtained must satisfy the initial distribution $\varphi(X)$ when $\mathcal{J} = 0$:

$$\varphi(X,\ 0) = p(X,\ 0) = \sum_{m=1}^{\infty} C_m \sin \frac{\pi m}{W_n} X, \qquad (3.35)$$

i.e., C_m are the coefficients of the Fourier function p(X, 0) when it is expanded as a series of sines in the range $0 < X < W_n$:

$$C_m = \frac{2}{W_n} \int_0^{W_n} p(X,\ 0) \sin \frac{\pi m}{W_n} X\, dX. \qquad (3.36)$$

Substituting into the above equation the values of p(X, 0) from Eq. (3.24) and integrating, we obtain the coefficients C_m.

As in the case of a diode with a thick base, the general solution (3.34) can be obtained also by the operator method and, in particular, by the two-dimensional Laplace-Carson transformation for a finite segment, which is widely used in calculations dealing with transient processes in thin-base diodes.

The special feature of all methods of finding the transient characteristics for a thin-base diode is that the solutions are always obtained in the form of an infinite series (in contrast to the integral form for a diode with a semi-infinite base). This makes it difficult to carry out a mathematical analysis and impossible to obtain general formulas in a sufficiently simple and clear form.

11.2. Transient Processes in a Diode with

Ohmic Recombination-Type Contact

Having found the value of C_m from Eqs. (3.36) and (3.24), we find that the general solution for the transient hole density distribution in a diode with a base of arbitrary thickness is

$$p(X,\ \mathcal{J}) = p_1 \sum_{m=1}^{\infty} \frac{2}{m\pi} \frac{\sin \frac{m\pi}{W_n} X}{1 + \left(\frac{W_n}{m\pi}\right)^2} \exp\left\{ -\left[1 + \left(\frac{m\pi}{W_n}\right)^2 \right] \mathcal{J} \right\}. \qquad (3.37)$$

The density of the transient reverse current is found, as usual, by differentiating $p(X, \mathcal{J})$ with respect to the position coordinate at $X = 0$:

$$j(\mathcal{J}) = 2j_f \frac{\tanh W_n}{W_n} \sum_{m=1}^{\infty} \frac{\exp\left\{-\left[1 + \left(\frac{m\pi}{W_n}\right)^2\right] \mathcal{J}\right\}}{1 + \left(\frac{W_n}{m\pi}\right)^2} . \qquad (3.38)$$

Equations (3.37) and (3.38) were derived first by Steele [12]. Let us analyze Eq. (3.38) in the limiting case $W_n \ll 1$. For a diode with a very thin base and not too low values of the time constant ($\mathcal{J} > W_n^2/\pi^2$), we need only keep the first term in the series represented by Eq. (3.38) and expand $\tanh W_n$ into a series. Then,

$$j(t) \simeq 2j_f \ \exp\left(-\frac{\pi^2}{W_n^2} \mathcal{J}\right) = 2j_f \ \exp\left(-\frac{\pi^2 D_p}{W^2} t\right), \qquad (3.39)$$

i.e., the reverse current decreases exponentially and its time constant is $W^2/\pi^2 D_p$, which is governed solely by the geometry of the base and is independent of the hole lifetime. It is important to note that this time constant is not equal to the effective lifetime introduced in the preceding section; this shows that the parameter τ_{eff} can be used only to estimate the stored charge and cannot be employed in analysis of all aspects of transient processes in a thin-base diode.

The functions $j(\mathcal{J})$ are shown graphically in Fig. 3.5 for several values of W/L_p. The curve for $W/L_p = \infty$ is plotted using Eq. (2.4) since the general solution (3.38) – which is valid for any value of W/L_p – cannot be used easily when $W/L_p \gtrsim 1$ because a very large number of terms in the series would have to be included. It is evident from Fig. 3.5 that when $W/L_p = 1$ the total duration of a transient process [defined as the time taken for the transient reverse current density $j(t)$ to decay to a value $0.1 \ j_f$ or close to that value] is approximately three times shorter than in the case of a long diode.

In the case of a thin base, we need retain only the first term in the series which gives the hole distribution [Eq. (3.37)] provided $\mathcal{J} > (W_n/\pi)^2$. Thus, a strongly asymmetrical initial distribution of holes (curve 1 in Fig. 3.4) transforms immediately after switching

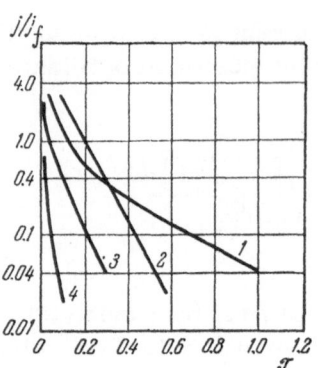

Fig. 3.5. Decay of the reverse current after the switching of diodes with different values of W/L_p and S_R in a circuit with $R_l = 0$: 1) $W/L_p = \infty$; 2) $W/L_p = 0.5$, $S_R = 0$; 3) $W/L_p = 1$, $S_R = \infty$; 4) $W/L_p = 0.5$, $S_R = \infty$.

into a symmetrical distribution representing a sinusoidal half-wave with the zero values at the boundaries (x = 0 and x = W) and a maximum in the middle of the base. The amplitude of this sinusoid decays with time but the symmetrical (with respect to the middle of the base) distribution of holes is retained throughout the transient process. Physically, this form of the function p(x, t) is explained by the fact that, after the application of a reverse voltage step to the p-n junction, the role of both planes (x = 0 and x = W) in the charge dispersal process is basically the same, since both of them are ideal absorbers of holes (although the mechanism of absorption is different for the p-n junction and for the ohmic contact). The stages of a transition from an initial asymmetrical hole distribution to a distribution in a diode with $W/L_p = 1$ are illustrated in Fig. 3.6. The shift of the maximum of the p(x,t) curve to the right, from x = 0 to x = W/2, takes place due to diffusion forces.

We shall now find the switching charge Q_{rec} for a thin-base diode in the absence of a limiting resistance in the circuit. Integrating term by term the series in Eq. (3.38) in the time interval

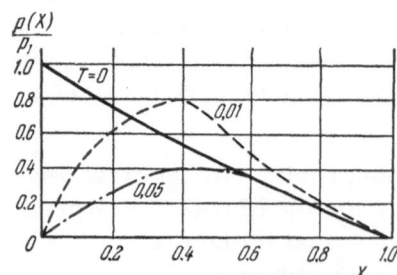

Fig. 3.6. Distribution of the density of excess holes at various times after switching in the base of a diode with $W/L_p = 1$ and $S_R = \infty$.

from 0 to ∞ and multiplying both sides of the equation by the p-n
junction area, we obtain

$$Q_{\text{rec}} = 2i_f \, \tau_p \frac{\tanh W_n}{W_n} \sum_{m=1}^{\infty} \left[2 + \left(\frac{W_n}{m\pi} \right)^2 + \left(\frac{m\pi}{W_n} \right)^2 \right]^{-1}. \qquad (3.40)$$

When $W_n < 1$, we obtain after neglecting the first two components
in the expression for the m-th term of the series in Eq. (3.40) and
using Eq. (3.21),

$$Q_{\text{rec}} \simeq \frac{1}{3} \frac{i_f \, W^2}{D_p} = \frac{2}{3} Q_{\text{st}}, \qquad (3.41)$$

since $\sum_{m=1}^{\infty} \frac{1}{m^2} = \frac{\pi^2}{6}$. Thus, when the base thickness of a diode with
a recombination-type ohmic contact is reduced, the amount of the
stored charge decreases, but the fraction of the charge leaking out
into the external circuit during the switching (the recovered charge)
increases somewhat (from $\frac{1}{2}$ for $W/L_p \rightarrow \infty$ to $\frac{2}{3}$ for $W/L_p \rightarrow 0$).
This is explained by the fact that, in contrast to a thick base, a
larger fraction of the charge stored in a thin-base diode is located
near the p-n junction.

11.3. Transient Processes in a Diode with a Noninjecting Ohmic Contact

The distribution of holes in the base of such a diode is given
by the solution of the diffusion equation (1.13) with the boundary
conditions (1.22), (3.5), and the initial condition (3.12), i.e., immediately after switching, the hole density near the p-n junction decreases to zero and there is no hole flux in the ohmic contact plane
at any time.

Using, as in § 11.1, the method of separation of variables, we
immediately obtain

$$p(X, \mathcal{T}) = \frac{2p_1}{W_n} \sum_{m=0}^{\infty} \frac{1}{\lambda_m} \frac{\exp\left[-\left(1 + \lambda_m^2 \right) \mathcal{T} \right]}{1 + \lambda_m^{-2}} \sin \lambda_m X. \qquad (3.42)$$

Differentiating Eq. (3.42) at $X = 0$ and using Eq. (3.13), we obtain
the following expression for the transient reverse current density:

$$j(\mathcal{J}) = 2j_f \, \frac{\coth W_n}{W_n} \sum_{m=0}^{\infty} \frac{\exp\left[-\left(1 + \lambda_m^2\right)\mathcal{J}\right]}{1 + \lambda_m^{-2}} \simeq \frac{2j_f}{W_n^2} \exp\left(-\frac{\pi^2 D_p}{4W^2}\,t\right), \quad (3.43)$$

where $\lambda_m = \pi(m + \frac{1}{2})/W_n$. The approximate equality in Eq. (3.43) is obtained on the assumption that $W_n = W/L_p \ll 1$ and $\mathcal{J} > (2W_n/\pi)^2$.

As in the case of an ohmic recombination-type contact, the reverse current decreases exponentially but the time constant is now four times larger than that in the case of $S_R = \infty$ (the pre-exponential term is $1/W_n^2$ times larger). These two differences indicate a much slower transient process in diodes with a noninjecting ohmic contact. Figure 3.5 shows the decay curve of the reverse current, calculated using Eq. (3.43) for a diode with $W/L_p = 0.5$. It is interesting to note that, during the initial phase of the transient process, the curve $j(\mathcal{J})$ lies higher than the curve for a diode with a semi-infinite base. This is due to a higher density of holes near the p-n junction, which is established during the flow of the forward current through a thin-base diode with a noninjecting ohmic contact [cf. Eq. (3.13)]. However, the decay of the reverse current is more rapid than in the case $W/L_p \rightarrow \infty$, and curves 1 and 2 intersect. This, in this case also we cannot find a time constant which would make it possible to describe the transient switching process in a thin-base diode using the analytic expressions obtained for a long diode.

For diodes with thin bases, we can neglect all terms in the series (apart from the zeroth term) of Eq. (3.42), provided $\mathcal{J} > (4W_n/\pi)^2$. In this case, the change in the hole distribution is opposite to that observed for a diode with $S_R = \infty$. A uniform distribution, which is observed before switching, very rapidly changes to strongly asymmetrical, representing a quarter of a sinusoidal wave whose zero value is located at $x = 0$ and whose maximum is at $x = W$. The nature of the distribution remains sinusoidal during subsequent stages of the decay but only the amplitude decreases.

Term-by-term integration of Eq. (3.43) with respect to time yields the following expression for the recovered charge:

$$Q_{\text{rec}} = 2i_f \, \tau_p \frac{\coth W_n}{W_n} \sum_{m=0}^{\infty} \left(2 + \lambda_m^2 + \lambda_m^{-2}\right)^{-1}, \quad (3.44)$$

which simplifies for $W_n \ll 1$ and assumes the form

$$Q_{rec} \simeq i_f \tau_p = Q_{st},$$ (3.45)

because $\sum\limits_{m=0}^{\infty} \dfrac{1}{(2m+1)^2} = \dfrac{\pi^2}{8}$. Equation (3.45) is physically self-evident: when the base thickness is reduced, the transient process of
the dispersal of the accumulated charge is so rapid that the influence of the recombination of holes in the base can be neglected;
the ohmic contact plane does not absorb holes. Therefore, the recovered charge is equal to the stored charge.

In §3, we have mentioned that the formulas discribing various transient processes in a diode with a semi-infinite base remain valid until $h_{p-n}/L_p \ll 1$, where h_{p-n} is the width of the space-
charge region In the analysis of thin-base diodes, this condition
has to be supplemented also by $h_{p-n}/W \ll 1$, which is more rigorous
when $W/L_p < 1$. However, if h_{p-n} is comparable with W, we then
observe the base-thickness modulation effect which is well known
in the transistor theory. The width of the space-charge region of
a p-n junction biased in the forward direction is considerably smaller than the width under a reverse bias. Therefore, in the expressions obtained by us for the forward-current case, we can use the
expression for W obtained earlier, but after switching the quantity
W must be replaced with $(W - h_{p-n})$, where h_{p-n} is the value at
the appropriate reverse voltage.*

11.4. Establishment of a Steady State after
the Application of a Forward Bias

Our analysis of transient processes in a thin-base diode applies to a steady-state hole distribution, i.e., it applies to conditions under which a given forward current is assumed to flow
through the diode for an infinitely long time before switching. To
obtain a criterion for the steady state and to consider the switching
of a thin-base diode under the conditions of "short" forward current
pulses, it is necessary to determine the law of establishment of a

---

* A rigorous solution of the problem of switching in a diode with a variable base thickness will not be given; the recommendations on the choice of the base thickness after
  switching are purely qualitative.

steady-state distribution of holes in the base. The solution of this
problem for a diode with a base of arbitrary thickness and with a
combination-type ohmic contact has been given by Baranov [84].
Using Eq. (1.13), or its equivalent Eq. (1.18), the initial distribution
of holes at zero time, and the boundary conditions [Eq. (3.9) for the
p-n junction and Eq. (3.2) for the ohmic contact plane], we find

$$p(X, \mathcal{J}_f) = \frac{j_f L_p}{q D_p} \left\{ \frac{\sinh(W_n - X)}{\cosh W_n} - \right.$$

$$\left. - \frac{2}{W_n} \sum_{m=0}^{\infty} \frac{\exp\left[-\left(1 + \lambda_m^2\right) \mathcal{J}_f\right]}{1 + \lambda_m^2} \cos \lambda_m X \right\}, \qquad (3.46)$$

where $\lambda_m = (m + \frac{1}{2})\pi/W_n$, and $\mathcal{J}_f$ is the dimensionless (normalized) duration of the forward current pulse.

Neglecting all the terms in the series (3.46) with the exception of the zeroth term (in the case of low values of W_n), expanding the hyperbolic sine and cosine in series, and taking only the
terms which are of the first order of smallness, we obtain the following approximate expression for the change in the hole density
near the p-n junction:

$$p(0, \mathcal{J}_f) \simeq p_1 \left[1 - \exp\left(- \frac{2.5}{W_n^2} \mathcal{J}_f\right)\right], \qquad (3.47)$$

which is accurate to within 10–15% when $W/L_p < 0.5$ and $p(0, \mathcal{J}_f)$
is not too small. To compare the processes of the establishment
of a steady-state hole distribution after the beginning of a forward

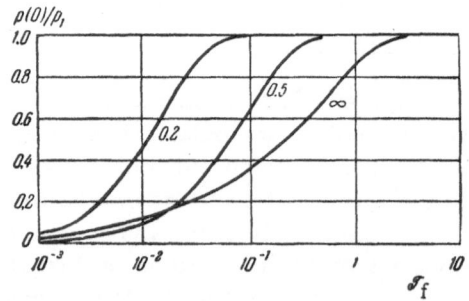

Fig. 3.7. Rise of the hole density to its impressed
equilibrium value during the application of a forward current pulse to diodes with bases of different thicknesses W/L_p.

current pulse in diodes with different base thicknesses, we have plotted in Fig. 3.7 the dependences given by Eq. (3.47) for $W/L_p = 0.5$ and 0.2, as well as the rise curve of the impressed hole density near a p-n junction in a diode with a semi-infinite base, plotted on the basis of Eq. (1.58). We must bear in mind that the steady-state hole density near the p-n junction is different for all three diodes, as shown by Eq. (3.9). The intersection of the curves corresponding to $W/L_p = \infty$ and $W/L_p = 0.5$ at low values of $\mathcal{J}_f$ is simply due to the inaccuracy of the approximation represented by Eq. (3.47). The thinner the base, the more rapidly is the value of p(0) reached (this value is represented in the figure as the ratio of the actual value and the maximum value). If the distribution of holes in the base is arbitrarily assumed to have reached a steady state when the value of p(0) reaches $0.9p_1$, the time taken to establish the steady-state distribution is $0.25\tau_p$ for $W/L_p = 0.5$ and less than $0.04\tau_p$ for $W/L_p = 0.2$.

Analysis of the establishment of the impressed hole density near a p-n junction in a diode with a base of arbitrary thickness, after the application of a forward current pulse, has been made also by Muratov and Kul'kin [85]. Numerical estimates based on the solution obtained by these authors show that, when $W/L_p = 2$, the impressed hole density at the p-n junction is established in practically the same manner as in a diode with an infinitely thick base.

When we consider the accumulation of excess holes in a thin-base diode with a noninjecting ohmic contact, we shall assume that at each moment the hole distribution in the base is uniform so that p(x, t) = p(0, t) at any point which satisfies 0 < x < W. The process of charge accumulation is then described by an ordinary differential equation

$$W \frac{dp(0)}{dt} = \frac{j_f}{q} - \frac{p(0) W}{\tau_p}, \qquad (3.48)$$

whose solution is

$$p(0) = p_1 \left(1 - e^{-\mathcal{J}_f}\right). \qquad (3.49)$$

Thus, the rise of the impressed hole density at the p-n junction has a time constant equal to τ_p, irrespective of the base thickness [provided the condition $W/L_p \ll 1$ is satisfied in the case of of Eq. (3.48)]. This result is not surprising because the total charge which should be accumulated in the base is equal, according to Eqs. (3.19) and (3.20), to $Q_{st} = i_f \tau_p$ and is also independent of the base thickness. The only process which prevents the hole density from reaching its equilibrium value p_1 is the recombination of holes in the base, described by the time constant τ_p.

A formula has been obtained by Muratov [86] for the transient reverse current in the case when switching takes place before a steady-state forward current is fully established in a thin-base diode with a recombination-type ohmic contact; however, this formula is so cumbersome that we shall not give it here. For a diode with a noninjecting ohmic contact, the transient reverse current is described by Eq. (3.43) even when the forward current has not had time to reach its steady-state value, but the quantity p_1 which occurs in j_f should be replaced by $p(0, \mathcal{J}_f)$ defined by Eq. (3.49).

§ 12. SWITCHING IN A CIRCUIT WITH A LIMITING RESISTANCE

We shall consider the transient process in a thin-base diode connected to a finite load resistance. As shown in § 5, in this case there is a time interval t_1 (known as the recovery phase during which the transient reverse current through the diode remains constant and depends only on the ratio U_r/R_l and not on the properties of the diode. The properties of the diode affect only the duration of the recovery phase, i.e., they determine the value of t_1. All that we have said in § 5 about the nature of the transient process during the recovery phase applies in full to thin-base diodes. The only change is in the actual form of the formulas.

To determine the duration of the recovery phase t_1, it is necessary to find a function $p(x, t)$, describing the distribution of holes in the base after switching, by solving the diffusion equation (1.13) using the boundary condition (1.20) in the p-n junction plane. The initial distribution of holes and the boundary condition in the non-rectifying contact plane should be taken in the form given by Eq. (3.8) and (3.2) or in the form of Eqs. (3.12) and (3.5), depending on the properties of this contact. The constant-reverse-current (recovery) phase ends when the density of holes at the p-n junction decreases to zero; from then onwards, the reverse resistance begins to increase and the reverse current begins to decay. Thus, when we find the function p(x, t), we can obtain an equation for the determination of the duration of the recovery phase in the form

$$p(0, t_1) = 0. \tag{3.50}$$

The distribution p(x, t) for the switching of a diode with a recombination-type ohmic contact in a circuit with a limiting resistance has been obtained by Byczkowski and Madigan [87], and since its form is similar to the distributions given by Eqs. (3.37) and (3.42), which have been considered in the preceding section, we shall not give them again. The substitution of the obtained function p(x, t) into Eq. (3.50) gives the following equation for the determination of t_1 for diodes with an arbitrary thickness of the base:

$$\frac{B}{1+B} \frac{W_n \tanh W_n}{2} = \sum_{m=0}^{\infty} \frac{\exp\left[-\left(1+\lambda_m^2\right)\mathcal{T}_1\right]}{1+\lambda_m^2}, \tag{3.51}$$

where $\lambda_m = \pi(m + \frac{1}{2})/W_n$, $\mathcal{T}_1 = t_1/\tau_p$, and $B = j_0/j_f$ is the switching parameter. For the most frequently encountered values of B, which are close to 1, Eq. (3.51) cannot be simplified because when W_n is reduced, $\mathcal{T}_1$ decreases as well and, therefore, at least several terms must be retained in the series. However, if B is small and $\mathcal{T}_1$ is large, we need retain only the zeroth term of the series for small values of W_n, expanding $\tanh W_n$ as a series; assuming that $\lambda_m^2 \gg 1$, we obtain an explicit expression for t_1:

$$t_1 \simeq 0.4 \frac{W^2}{D_p} \ln 0.8 \frac{1+B}{B}, \tag{3.52}$$

which, as estimated numerically in [88], is valid when $t_1 \geq 0.1$ W^2/D_p.

For one of the special cases, B = 1, the solution of the general equation (3.51) is shown graphically in Fig. 3.8. Examination of curve 1 in Fig. 3.8 shows that when $W/L_p \geq 2$, we may assume that the diode has a semi-infinite base. At low values of W_n the duration of the recovery phase decreases proportionally to W_n^2,

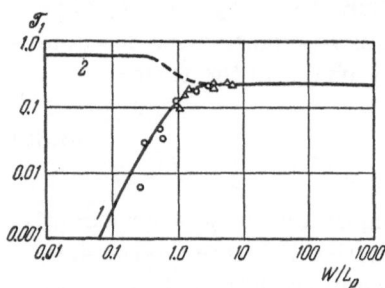

Fig. 3.8. Dependence of the duration
of the first (recovery) phase on the base
thickness for diodes with $S_R = \infty$ (1)
$S_R = 0$ (2) for $B = 1$. The dashed part
of curve 2 in the region $0.5 < W_n < 2$
was not calculated but interpolated.
The experimental points shown in the
figure were obtained for two p-type
silicon plates.

which is in agreement with the dependence of $\mathcal{J}_1$ on W_n given by
the approximate formula (3.52) for low values of B.

The decay of the reverse current after the end of the recovery
phase can be found by solving the diffusion equation with a new
boundary condition for the p-n junction: $p(0, t) = 0$ when $t > t_1$; as
the initial distribution we can use the function $p(x, t_1)$. However,
the distribution of holes in the base at a time $t = t_1$ for the case $B \geq 1$
is close to the distribution which would be obtained by switching a
diode in a circuit without a limiting resistance at a time $t = t_1'$,
where $j(t_1') = j_0$. Therefore, in the first approximation we may assume that the decay of the reverse current after the end of the recovery phase is described by Eqs. (3.38) and (3.39) provided the
time is measured from $t = t_1'$, defined by the relationship $j(t_1') = j_0$.

To estimate the relative importance of the first (recovery)
and second (reverse) phases of a transient process, we shall introduce the concept of the rate of recovery of the reverse current
(or the reverse resistance) of the diode. We shall arbitrarily assume the second phase of the recovery to be complete at a time t_2
from its beginning when the reverse current decreases to $i_2 = 0.1\, i_0$.
The ratio $\beta = t_2/t_1$ is a quantitative measure of the rate of recovery
of the reverse current. The lower the value of β, the greater is the
importance of the first (recovery) phase and the closer the form
of the transient characteristic of the diode to the rectangular shape
and the more rapid is the recovery.* Using the approximate expression (3.39) and Fig. 3.8, we find that for $B = 1$ and $W_n < 1$, diodes

---

* Recently, in connection with the development of charge-storage diodes (cf. for example [62, 103]) and in connection with the search for diodes with a retangular transient
characteristic, interest in the rate of recovery is increasing.

with a recombination-type ohmic contact obey

$$\beta = t_2/t_1 \simeq 1.1 \tag{3.53}$$

and the value of β given by the above equation is independent of the base thickness. An estimate of the rate of recovery of planar diodes with semi-infinite bases, obtained using Figs. 2.12 and 2.13, gives $\beta \approx 2.5$ for B = 1. Thus, for diodes with a recombination-type ohmic contact, a reduction of the base thickness from W = ∞ to W $\leq 0.5 L_p$ produces an approximately twofold increase of the rate of recovery of the reverse current. Moreover, the total duration of the transient process also decreases.

In the case of a noninjecting ohmic contact, when the distribution of holes in thin bases can be regarded as uniform at any given moment [this follows from Eq. (3.12)], the dispersal of the accumulated charge is described by an ordinary differential equation of the (3.48) type provided we substitute j_0 for j_f. The expression obtained in this way for the duration of the first (recovery) phase of the switching process, taking into account only the recombination in the base and neglecting the diffusion processes, is of the form

$$t_1 = \tau_p \ln \frac{1+B}{B} = \frac{Q_{st} B}{i_0} \ln \frac{1+B}{B}, \tag{3.54}$$

i.e., for thin bases ($W_n < 0.5$) the value of t_1 is independent of W and is governed only by the hole lifetime and switching conditions. The second equation in Eq. (3.54) shows that t_1 is the time of the dispersal of the charge Q_{st} stored in the base by the flow of the reverse current i_0, which takes into account partial recombination of this charge in the base.

For the sake of comparison with the conditions just discussed, Fig. 3.8 gives the value of t_1 for B = 1; the dashed part of the curve in the range $0.5 < W_n < 2$ has not actually been calculated but extrapolated. Comparison with a planar diode having a semi-infinite base is given in Fig. 3.9. It is evident from this figure that at high values of B the difference between the durations of the recovery phase for diodes with very thin and very thick bases (having the same values of τ_p) may be very considerable.

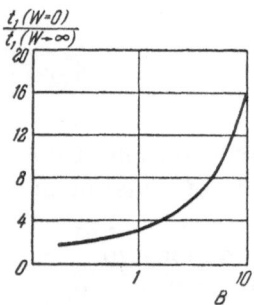

Fig. 3.9. Dependence, on the parameter B, of the ratio of the recovery phase durations for diodes with very thin and very thick bases and $S_R = 0$.

In agreement with the assumptions made in the derivation of Eq. (3.48), the excess hole charge decreases to zero at the end of the first (recovery) phase. This means that the reverse current decreases instantaneously from i_0 to zero, i.e., for a diode with a noninjecting ohmic contact we have $\beta \approx 0$. In practice, the duration of the second phase t_2 is always finite but when W_n is reduced, this duration can be reduced to a very small value. Assuming that t_1 remains constant, we may expect to obtain values $B \ll 1$ for thin-base diodes with noninjecting ohmic contacts ($S_R = 0$).

The formula (3.51) has been checked experimentally by Byczkowski and Madigan [87] for silicon planar diodes. Rectifying junctions were prepared by the high-temperature diffusion of phosphorus (or boron) into a thick p-type (or n-type) silicon plate. Prolonged annealing followed by slow cooling made it possible to maintain the minority carrier lifetime at a level of several microseconds in such p-n junction structures. Mesa structures were produced by etching on the p-n junction side. The opposite side of the plate was ground to remove the diffused layer and then electroplated with a film of nickel which acted as a recombination-type ohmic contact (Fig. 3.10). After the measurement of the duration of first phase t_1 under the selected switching conditions, the plates were ground in a controlled manner on the ohmic contact side, nickel was deposited again, and the value of t_1 was remeasured. Such repeated grinding away of the silicon layers made it possible to determine experimentally the dependence $t_1 = f(W)$. The minority carrier lifetime was determined from the duration of the first phase for a diode with a very thick base ($W_n > 5$) using Eq. (2.28). The presence of a large number of noninteracting p-n junctions in one plate made it possible to carry out a statistical analysis of the results. Experimental points for two p-type silicon plates of 15 and 25 $\Omega \cdot$cm resistivity and with electron lifetimes of 3 and 6 μsec, plotted in Fig. 3.8, confirmed that the agreement between experiment and theory was good.

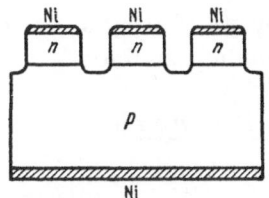

Fig. 3.10. Structure of diffused p-n junctions.

The duration of the first (recovery) phase in a thin-base diode for a nonlinear boundary condition in the ohmic contact plane was determined by Tkhorik [82]. He assumed that the value of S_R was not constant but depended on the carrier density near the nonrectifying contact in accordance with the law described by formula (3.15) and by the linear part of the curve in Fig. 3.2. The solution was obtained on the assumption that during the time t_1 the carrier density at $x = W$ did not change greatly, i.e., the boundary condition for this plane was of the form (3.1) with a constant right-hand side. The nonlinearity was thus taken into account only in the initial distribution of holes in the base. The solution obtained under these assumptions is shown in Fig. 3.11 in the form of the dependences of $\mathcal{J}_1$ on j_f. When the density of the forward current increased due to a reduction in the surface recombination velocity in the ohmic contact plane (Fig. 3.2), the charge stored in the base increased superlinearly with j_f. This was responsible for the increase in the response time of the diode when j_f was increased but B was kept constant. This effect has been observed experimentally by Tkhorik [82] in an investigation of a batch of diffused germanium diodes with ohmic

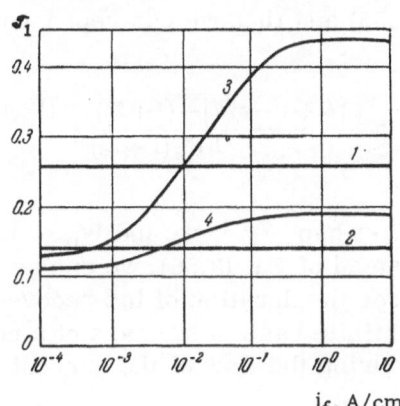

Fig. 3.11. Dependence of the duration of the first (recovery) phase on the forward current density. 1) $W_n \geq 3$, B = 1; 2) $W_n \geq 3$, B = 2; 3) $W_n = 1$, B = 1; 4) $W_n = 1$, B = 2.

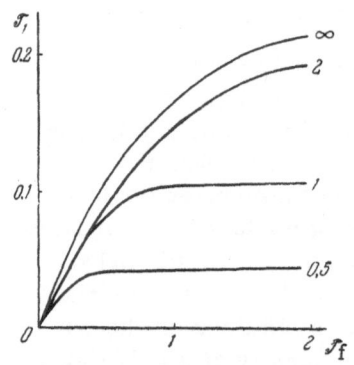

Fig 3.12. Dependence of the duration of the first (recovery) phase on the duration of the forward current pulse for diodes with bases of different thicknesses W_n.

contacts prepared by tin-plating a groung p-type germanium surface.

12.1. Short Forward Current Pulses

If, before switching, the forward current flows through a diode for a finite time, it is necessary to use the nonstationary distribution of holes given by Eq. (3.46) as the initial condition. The equation for $\mathcal{J}_1$, obtained for this case using the boundary conditions (1.20) and (3.2), gives [86, 88]:

$$\frac{BW_n \tanh W_n}{2} = \sum_{m=0}^{\infty} \{B + 1 - \exp\left[-\left(1 + \lambda_m^2\right)\mathcal{J}_f\right]\} \frac{\exp\left[-\left(1 + \lambda_m^2\right)\mathcal{J}_1\right]}{1 + \lambda_m^2}, \quad (3.55)$$

where λ_m has the same value as in Eq. (3.51). We can easily see that when $\mathcal{J}_f \to \infty$, Eq. (3.55) transforms to Eq. (3.51). For relatively large values of t_1 [they can be regarded as large under the conditions applying in the case of Eq. (3.52)], we need retain only the zeroth term in Eq. (3.55) and then we can represent $\mathcal{J}_1$ explicitly in the form:

$$\mathcal{J}_1 = \frac{1}{1 + \lambda_0^2} \ln \frac{2\{B + 1 - \exp\left[-\left(1 + \lambda_0^2\right)\mathcal{J}_f\right]\}\coth W_n}{BW_n\left(1 + \lambda_0^2\right)}, \quad (3.56)$$

which reduces to (3.52) when $\mathcal{J}_f \to \infty$ and $W_n \ll 1$. Numerical estimates show that, instead of Eq. (3.56), we can use a similar formula (2.36), which gives the duration of the recovery phase t_1 for a diode with a semi-infinite base in the case of a nonstationary distribution of holes during the flow of the current when $(\mathcal{J}_1 + \mathcal{J}_f) < 0.4W_n$.

It is evident from the structure of Eq. (3.55) that a forward current pulse can be regarded as infinitely long if the term containing $\mathcal{J}_f$ (in braces) becomes much smaller than $(1 + B)$. Since the minimum value is $B = 0$, we can assume the forward current pulse to be infinitely long when

$$\exp\left[-\left(1+\lambda_m^2\right)\mathcal{J}_f\right]<0.1.$$

Hence, after some transformations, we obtain the following estimate for the minimum value of t_f, for a thin-base diode ($W_n < 1$), above which we can assume the forward current pulse to be infinitely long:

$$t_f \gtrsim \frac{W^2}{D_p}. \tag{3.57}$$

Figure (3.12) shows the dependence of the duration of the first (recovery) phase $\mathcal{J}_1$ on the duration of the forward current pulse $\mathcal{J}_f$ before switching, calculated using Eq. (3.55) for the case B = 1. The curve for $W_n = 0.5$ confirms the validity of the estimate represented by Eq. (3.57); comparison of the dependences $\mathcal{J}_1 = f(\mathcal{J}_f)$ for diodes with $W_n = 2$ and $W_n = \infty$ shows that when $W_n = 2$ the diode behaves practically as a diode with a semi-infinite base.

When a thin-base diode has a noninjecting ohmic contact ($S_R = 0$), the duration of the first phase t_1 under transient conditions is still given by (3.54), provided we substitute in that formula $Q_{st}(\mathcal{J}_f) = Q_{st}(\mathcal{J}_f \to \infty)\left(1 - e^{-\mathcal{J}_f}\right)$. Thus, in this case, the forward current current pulse can be assumed to be infinitely long only when $t_f > (2\text{-}3)\tau_p$, irrespective of the base thickness.

12.2 Postinjection Voltage

The decay of the postinjection voltage across a p-n junction, observed after the end of the forward current pulse under open-circuit conditions in thin-base diodes, is basically similar to the corresponding decay for diodes with semi-infinite bases. The presence of a recombination-type contact in the vicinity of a p-n junction accelerates the leakage of holes from the base and, therefore, it accelerates the transient process in the open circuit. The closeness of a reflecting boundary (a contact with $S_R = 0$) to a p-n junction prevents the diffusion leveling of the hole density in the base, which should slow down somewhat the decay of the postinjection emf during the initial stage of the process, when the role of the diffusion is considerable (cf. §6). After a sufficiently long time, the decay of the voltage "tail" should, as in the case $W \rightarrow \infty$, be described by the time constant τ_p.

The general solution for the transient process of the post-injection emf decay in a diode with an arbitrary base thickness and an arbitrary surface recombination velocity in the ohmic contact plane was obtained by Gaman and Kalygina [89]. They used the diffusion equation for a low injection level (1.13), the steady-state initial distribution of holes of the type given by Eq. (3.6), and the boundary conditions (3.1) and (1.20) with $j = 0$. The solution of $p(0, \mathcal{T})$ was obtained in the form of a series with a very cumbersome general term. For sufficiently large values of $\mathcal{T}$, only the first terms of this series had to be retained and, therefore, it was possible to obtain an explicit expression for the description of the decay u(t). It was found that, as in the case of a thick-base diode, there was a linear part of the postinjection voltage decay curve which satisfied the following relationship:

$$\frac{du(t)}{dt} = -\frac{kT}{q}\frac{1}{\tau_w}, \qquad (3.58)$$

which was equivalent to (2.52) in which τ_p was replaced with τ_w. The value of τ_w was found to depend on the hole lifetime in the base, the base thickness, and the recombination velocity at the ohmic contact:

$$\frac{1}{\tau_w} = \frac{v_0^2 D_p}{W^2} + \frac{1}{\tau_p}, \qquad (3.59)$$

where v_0 is the first root of the transcendental equation

$$\cot v_0 = \frac{v_0 D_p}{W S_R}. \qquad (3.60)$$

We shall consider two extreme (limiting) cases. When $S_R = \infty$, $v_0 = \pi/2$, and, correspondingly, $1/\tau_w = 2.5 \, D_p/W^2 + 1/\tau_p$; however, when $S_R = 0$, $v_0 = 0$, and $\tau_w = \tau_p$. For sufficiently thin bases ($W/L_p < 0.5$), the time constant for a diode with $S_R = \infty$ becomes

$$\tau_w \simeq 0.4 \frac{W^2}{D_p}. \qquad (3.61)$$

An estimate obtained for a high injection level in a diode with a recombination-type ohmic contact [55] shows that during the initial stage of the transient process the decay of u(t) is again linear and the slope is $\frac{2b}{b+1} \frac{kT}{q} \frac{1}{\tau_W}$, but when the impressed density of holes decrease to values smaller than n_{n0}, the decay curve obeys Eq. (3.58).

On approach to the equilibrium state [u(t) ≪ kT/q], the linear decay of the voltage across the p-n junction changes to exponential but the time constant τ_W remains the same.

Thus, the decay of the postinjection emf under open-circuit conditions in thin-base diodes is exactly the same (in the most interesting region, i.e., some time after the end of the forward current pulse) as in the case of a thick-base diode, except that the hole lifetime τ_p in the formulas is replaced by a time constant τ_W, which depends on τ_p, W, and S_R.

§ 13. GENERAL ESTIMATE OF THE RESPONSE OF A THIN-BASE DIODE

It is desirable now to consider some generalizations of the results obtained in the analysis of transient processes in thin-base diodes.

First of all, we find that when the base thickness is reduced, the properties of the nonrectifying contact become decisive in the response of a diode. If this contact is of the recombination type ($S_R = \infty$), the diode response time decreases when W is reduced (this happens under all switching conditions); however, if this contact is noninjecting ($S_R = 0$), the response time increases under all switching conditions with the exception of the reverse current decay at high values of t.

For diodes with both types of contact, the base can be regarded as infinitely thick when

$$W > 2L_p \tag{3.62}$$

and very thin when

$$W < 0.5L_p. \tag{3.63}$$

The first relationship follows from the analysis of the dependence of the stored charge Q_{st} on W given by Eq. (3.19), from the nature of the process of establishment of the impressed hole density during the application of a forward current pulse (cf. §11), and from the dependence of the duration of the first (recovery) phase on the base thickness during the switching from the steady and nonsteady states (cf. Figs. 3.8 and 3.12).

The validity of the inequality (3.63) is confirmed by analysis of the general term of the series in Eqs. (3.37), (3.38), (3.42), (3.43), (3.46), and (3.51). In all cases, when W < $0.5L_p$, we find that $\lambda_m^2 \geq 10$ and, therefore, we can neglect unity compared with λ_m^2 so that we obtain the corresponding simplified expressions for diodes with very thin bases.

The transient processes in thin-base diodes cannot be, in general, reduced to the case of a diode with a semi-infinite base by introducing some new time constant, similar to the hole lifetime τ_p for a diode with W → ∞. However, some conclusions can still be drawn.

The duration of practically all the transient processes taking place during the switching of a diode with a noninjecting ohmic contact ($S_R = 0$) are proportional to the value of τ_p, irrespective of the base thickness. This is because the stored charge is $Q_{st} = i_f\tau_p$ and hole recombination does not take place in the ohmic contact plane. Only the decay of the reverse current in a circuit without a limiting resistance is governed (in its final stage) by the transit time of holes through the base, i.e., by the term which depend on W^2/D_p.

The duration of all the transient processes in a thin-base diode with a recombination-type ohmic contact ($S_R = \infty$) is governed only by the transit time through the base. When an effective leakage of excess holes takes place through both boundaries of the base (this happens, for example, during switching in a circuit without a limiting resistance), the time constant is $W^2/\pi^2 D_p$; in other cases, when the role of the p-n junction in the dispersal of holes is much less important then the role of the ohmic contact, the time constant is correspondingly larger and equal to $4W^2/\pi^2 D_p$. In all the expressions describing the transient processes in thin-base diodes, we always encounter the ratio W^2/D_p.

An experimental check of the principal calculated relationships for thin-base diodes has been carried out less thoroughly than for thick-base diodes. This is primarily because the preparation of nonrectifying contacts with a given value of S_R or even the determination of the value of S_R of an existing contact meets with very considerable, and sometimes insurmountable, experimental difficulties. In fact, apart from the experimental work of Byczkowski and Madigan, [87] as well as the experiments of Penin and Cherkas on small-signal frequency characteristics of thin-base diodes [90], other investigators [31, 83] have concentrated not so much on the checking of the theory as on the determination of the electrical characteristics by reduction of the diode base thickness.

The theory of transient processes in thin-base diodes has been developed less thoroughly than the theory for long (or thick-base) diodes. This is because the mathematical analysis of diodes with finite values of W is very difficult. Moreover, the limited possibilities of experimental verification have also retarded the development of the theory. Finally, because thin-base diodes with non-injecting contacts do not show much promise in fast-response pulse device applications, the theory of such devices has been paid little attention. In view of the very rapid development of charge-storage diodes, which can be thin-base diodes, further development of the theory of transient processes in thin-base diodes is an urgent problem.

Chapter IV

Transient Processes in a Diode with a
Small-Area Rectifying Contact

The earliest experimental investigations of transient processes in semiconductor devices were carried out on point-contact
germanium diodes [91-95]. The reduction of the reverse resistance under pulse conditions and the decrease of the rectified current with increasing frequency were observed for these diodes.
The first explanations of the experimentally observed strong peaks
of the reverse current correctly attributed these peaks to the injection of holes into the base during the flow of the forward current
and the return of these holes to the contact when a reverse voltage
was applied. Meacham and Michaels [92] introduced the term
"charge storage" to describe an increase in the hole density in the
base due to the injection of holes by the p-n junction.

A theory of the processes taking place during the switching
of a diode with a planar p-n junction accounts qualitatively for the
relationships observed experimentally in the switching of point-
contact diodes. However, a quantitative agreement between theory
and experiment is not obtained: the transient processes in point-
contact diodes are considerably faster than could be expected from
the hole lifetime in the original germanium material.

The difference between theory and experiment is due to the
fact that real point-contact diodes cannot be described by a planar
diode model. The most important difference is that the rectifying
contact in a point-contact diode cannot be regarded as infinite
since its linear dimensions are frequently comparable with the diffusion length of holes in the base and this diffusion length is a characteristic dimension governing the volume of the semiconductor base
in which the effective accumulation and dispersal of holes take place.

Therefore, it has been necessary to carry out special theoretical calculations of transient processes in point-contact diodes or, in a more general sense, in diodes with small-area rectifying contacts.

§ 14. IDEAL MODEL OF A POINT-CONTACT DIODE

14.1. Structure of a Point-Contact Diode

The majority of point-contact diodes are prepared by pressing a sharp tungsten wire against the surface of an n-type germanium crystal and "electroforming," which consists of passing of a sufficiently powerful electric pulse through the contact.

As a result of the electroforming process and the associated evolution of a considerable amount of heat under the point contact, a small (of $\sim 10^{-3}$ cm radius) p-type germanium region is formed under the contact. This produces a p-n junction in the contact region. Numerous investigatigations have established that the injection efficiency of electroformed point contacts is considerably less than unity, but it can be fairly high. Investigations of the geometry of the electroformed p-type region [96, 97] have shown that its shape resembles a pear with the thicker end in the semiconductor and the thinner end at the point of emergence of the p-type region on the surface. Under some electroforming conditions, the p-type region may be hemispherical. The generation of thermal acceptors during electroforming is due to the diffusion of copper atoms and, therefore, the majority carrier density in the p-type region is usually close to the maximum solubility of copper in germanium, which is 3×10^{16} cm^{-3}.

In some cases, an acceptor impurity film (usually indium or aluminum) is deposited on the tip of the contact point; this increases the density of holes in the p-type region to 10^{17} cm^{-3} or more, and the value of the injection efficiency is raised to $\gamma > 0.9$-0.95. This process also increases the p-type region.

Another widely used method of preparing small-area rectifying contacts consists of welding thin wires, made of materials which form a low-temperature eutectic alloy with germanium (usually gold or silver), to a germanium crystal. When a small admixture of gallium is added to the material of such a wire, the concentration of acceptors in the recrystallized p-type germanium layer may exceed 10^{19} cm^{-3}. Thus, the injection efficiency in these gold or silver point-contact diodes is practically equal to unity. Investigations of the geometry of the p-n junctions in such diodes [98, 99] have shown that, in all cases, the p-type region is nearly hemispherical. The diameter of the alloyed region in some types of welded point-contact diode does not exceed 10 μ.

In view of this, in any theoretical analysis of transient processes in diodes with small-area rectifying contacts, it is usual to employ the following ideal model (Fig. 4.1) In a semi-infinite n-type semiconductor we have a p-type conduction region in the form of a hemisphere of radius r_0 and it is assumed that the injection efficiency γ of the resultant p-n junction is equal to unity. The lines of the current flow are assumed to be radial, which is correct only if the recombination of carriers on the surface of the n-type region can be neglected, i.e., if $S_R = 0$. The ohmic contact is assumed to be infinitely far away from the p-n junction so that its properties do not affect the characteristics of the diode. For simplicity, we shall call the structure shown in Fig. 4.1 a model of a diode with a hemispherical p-n junction, a point-contact diode model, or (in those cases where not confusion will arise) a point-contact diode.

Naturally, the results of the calculations carried out for this model describe accurately only the transient processes in a gold-of silver-welded diode, as well as the time dependences of contributions to the transient reverse current due to the accumulation of holes in the base of an electroformed point-contact diode.

Usually, the exact shape of the surface of the rectifying contact is not known. Then, when we compare theory with experiment we proceed as follows: we find the p-n junction area by measuring the barrier capacitance of the diode and then, assuming that the

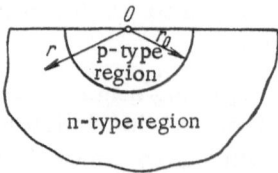

Fig. 4.1. Model of a diode with a hemispherical p-n junction.

junction is hemispherical, we determine the effective radius of the hemisphere from $r_0 = \sqrt{S/2\pi}$, which is then used in theoretical formulas.

14.2. Solution of the Diffusion Equation

for the Steady-State Case

Since a point-contact diode is inhomogeneous, it is necessary to use the general diffusion equation which is of the following form [for the same assumptions which have been used to derive Eq. (1.13)]:

$$\nabla^2 p\,(x,\,y,\,z,\,t) - \frac{p\,(x,\,y,\,z,\,t) - p_{n0}}{L_p^2} - \frac{1}{D_p}\,\frac{\partial p\,(x,\,y,\,z,\,t)}{\partial t} = 0, \qquad (4.1)$$

where ∇^2 is the Laplace operator. In the spherical symmetry case, when $p_{n0} = 0$, in terms of dimensionless variables $\mathcal{J} = t/\tau_p$ and $R = r/L_p$, Eq. (4.1) becomes

$$\frac{\partial p\,(R,\,\mathcal{J})}{\partial \mathcal{J}} + p\,(R,\,\mathcal{J}) = \frac{\partial^2 p\,(R,\,\mathcal{J})}{\partial R^2} + \frac{2}{R}\,\frac{\partial p\,(R,\,\mathcal{J})}{\partial R}. \qquad (4.2)$$

This equation describes the behavior of holes at any moment for all values of $R \geq A$, where $A = r_0/L_p$. When we neglect the term p_{n0} in Eq. (4.2), we imply that the reverse current is, in all cases, much greater than the saturation current. The boundary conditions for the planar model, given by Eqs. (1.20), (1.21), (1.22), reduce to the following form in the spherical geometry case: for a constant forward current through the p-n junction, we have

$$\left(\frac{\partial p}{\partial R}\right)_{R=A} = -\frac{jL_p}{qD_p}; \qquad (4.3)$$

for a constant voltage across the p-n junction, we obtain

$$p_1 = p_{n0}\exp\frac{qu_{p-n}}{kT}; \qquad (4.4)$$

or, when the reverse voltage is sufficiently large $|u_{p-n}| \gg kT/q$,

$$p(R = A,\,\mathcal{J}) = 0. \qquad (4.5)$$

Here, p_1 represents the density of holes impressed by the p-n junc-junction at R = A.

These relationships are supplemented by the self-evident con-dition that $p \to 0$ when $R \to \infty$, which means that very far from the p-n junction there is no accumulation of holes. Strictly speaking, $p = p_{n0}$ when $R \to \infty$ but, in fact, we have assumed that $p_{n0} = 0$ when $R \to \infty$. The steady-state distribution of holes in the diode base during the flow of a constant forward current is found by solving Eq. (4.2) (assuming that $\partial p / \partial \mathcal{J} = 0$) with the boundary condition given by Eq. (4.3); the solution is

$$p = p_1 \frac{A}{R} e^{-(R-A)}. \tag{4.6}$$

Hence, from Eq. (4.3), we find the relationship between the forward current density and the impressed hole density at the p-n junction:

$$j_f = \frac{qD_p p_1}{L_p} \left(\frac{A+1}{A} \right) \simeq \frac{qD_p p_1}{r_0}. \tag{4.7}$$

The approximate equality in (4.7) applies to diodes with a very small radius of the p-type region ($r_0 \ll L_p$). It is interesting to note that the decrease of the hole density away from the p-n junction in point-contact diodes is more rapid than in planar diodes be-cause Eq. (4.6), which gives the value of p, contains an exponential term as well as a factor 1/R. The expression (4.7), relating j_f and p_1, is fully analogous to Eq. (1.26) for a planar diode except that the characteristic length is now r_0 and not L_p.

Integration of Eq. (4.6) over the whole volume of the base gives the value of the stored charge. We can easily show that

$$Q_{st} = i_f \tau_p, \tag{4.8}$$

i.e., this expression is exactly the same as for a planar diode, and is independent of the radius of the p-n junction. However, the dis-tribution of the excess hole charge in the base of a point-contact diode differs appreciably from the distribution in a planar diode.

Figure 4.2 shows the radial distribution of the charge stored in a spherical layer of thickness dr, separated by a distance r from the center of the p-type region. The dependences shown in Fig. 4.2 are plotted for diodes with various values of the ratio A, including the ratio $A = \infty$ for a planar diode; it is assumed that the value of the forward current is the same in all cases. The excess charge density is found from the relationship $dQ_{st} = qp(r) 2\pi r^2 dr$. The curves plotted in Fig. 4.2 show clearly the distribution of the charge in the base: the fraction of the charge in those regions of the base which are not further than a distance r' from the p-n junction is proportional to the area under the corresponding curve up to the coordinate r = r'. Examination of the curves in Fig. 4.2 shows that, when the radius of the p-n junction is reduced, the fraction of the charge accumulated near the junction and in the region extending up to $r = r_0 + L_p$ decreases. Such a distribution of the excess hole charge prevents us from improving the diode response by reducing the parameter r_0. Moreover, when we reduce the rectifying contact area, we increase the hole density gradient near the p-n junction. In fact, it follows directly from Eq. (4.3) that

$$\left(\frac{\partial p}{\partial R}\right)_{R=A} = -\frac{i_f L_p}{2\pi q D_p} \frac{1}{r_0^2}, \qquad (4.9)$$

i.e., the diffusion flow of holes away from the p-n junction into the interior of the base is proportional to $1/r_0^2$. It follows from Eq. (4.9) that as $r_0 \to \infty$, i.e., as we go over to a planar diode, we find that $\partial p/\partial R \to 0$. There is no contradiction in this conclusion since we have assumed a constant value of the forward current for all the diodes considered; naturally, when the junction area is increased the current density and the hole density gradient decrease.

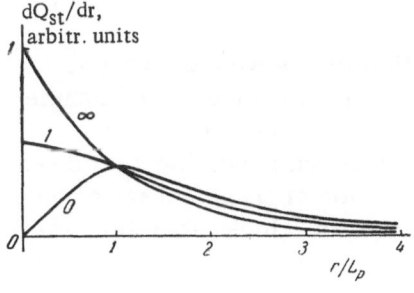

Fig. 4.2. Distribution of the stored charge in the base of a point-contact diode for various values of A which are given alongside the curves (r is measured from the p-n junction).

Thus, the characteristic features of the spherical geometry of a p-n junction give rise to two main differences between point-contact and planar diodes, which are important in comparison of their response.

When the same constant forward current passes through both diodes, the charge stored in a point-contact diode is distributed in regions of the base further from the p-n junction and the hole density gradient near the p-n junction is greater than in the planar case.

14.3 Refinement of the Basic Assumptions

Because of the small p-n junction area and the related high forward current density in point-contact diodes, the high injection level condition applies in practically all cases of switching in such diodes. We have demonstrated in Chaps. I and II that the driving field, which appears when $\Delta \gg 1$, produces a very slight "pulling" of holes into the base. Comparison of the values of L_p and of the "effective diffusion length" l_p at high injection levels shows that, in the first approximation, the influence of the driving field is equivalent to a change of the diffusion coefficient from D_p to $2D_p D_n/(D_p + D_n)$.

When a p-n junction is fabricated in a point-contact diode, a considerable number of structure defects is generated in the contact region and these defects reduce the values of D_p and D_n compared with the values for a pure and perfect semiconductor. If we assume that the concentration of defects produced by electroforming is close to 10^{17} cm^{-3}, we find (using curves similar to those shown in Fig. 2.17) that $2D_p D_n/(D_n + D_p) \simeq 45$ cm^2/sec, i.e., the effective diffusion coefficient is practically equal to the value of D_p in pure germanium.

It follows that even at high injection levels we can use Eq. (4.2) without committing an appreciable error, provided we assume that the hole lifetime is independent of the density of excess carriers. The results obtained using Eq. (4.2) and assuming $\tau_p = $ const give somewhat greater values of the duration of transient processes than would be obtained by solving the equation which takes into account the drift of holes.

Because the concentration of acceptors in the p-type region of some types of point-contact diode is low, the injection and accumulation of electrons in the p-type region in these devices becomes important even at relatively low values of the forward current. However, since the dimensions of the p-type region are very small, the dispersal of electrons may affect only the initial stage of the transient process but not the final stage, which is governed by the dispersal of holes from the base. On the other hand, because of the reduction of the potential barrier in the region of the p-n junction and because of the decrease in the injection efficiency, we should observe a relative decrease of the response time of a diode when the forward current is increased and this should be observed for values of i_f and j_f much smaller than those at which the response times decreases in planar diodes (cf. §5).

The condition that the space-charge region width h_{p-n} should be small compared with the diffusion length of holes L_p (this must be satisfied in the transient processes in planar diodes) is replaced by a more rigorous condition in the case of point-contact diodes:

$$h_{p-n} \ll r_0.$$

Only when this condition is satisfied can we assume r_0 to be the parameter of the diode having the same value for the forward and reverse voltages across the p-n junction.

Usually, the dimensions of the rectifying contact are reduced as much as possible in the production of high-frequency and fast-response pulse diodes. Another measure necessary for the increase of the speed of response of diodes is the reduction of the resistivity of the initial semiconducting material (only by reducing ρ can we retain a sufficiently low value of the spreading resistance of a point contact when we reduce its area). The thickness of the space-charge region of the p-n junction also decreases when ρ is reduced. Consequently, in the majority of point-contact electroformed and welded diodes we can neglect the transient phenomena in the depletion layer and we may assume r_0 to be a constant parameter of a given diode.

In real point-contact diodes the ohmic contact is not an infinite distance from the p-n junction. However, since the processes of accumulation and (as we shall show later) dispersal of holes are, to a great extent, governed by the value of the p-n junction radius and not by the diffusion length, it follows that we can compare the value of W with the parameter r_0, and not with L_p, in considering whether the base can be regarded as semi-infinite.

The practical experience in the production of point-contact diodes shows that, in all cases, a satisfactory current-voltage characteristic of an electroformed or welded contact is obtained only when the thickness of a crystal is not too small. The usual relationship is $(W/r_0) > 1$ (here, W is understood to be the minimum distance between the rectifying and ohmic contacts). Therefore, even when $(W/L_p) \lesssim 1$, we are justified in assuming the base of a point-contact diode to be semi-infinite provided the relationship $(W/r_0) > 1$ or, which is even better, $(W/r_0) \geq 2\text{-}3$ is satisfied.

14.4. Role of Surface Recombination

We shall now consider the conditions of the validity of the assumption that the flow lines of the current in the base are radial. We shall assume that the surface of germanium can be represented by a surface recombination velocity S_R, which is the same for any distance from the p-n junction. In this calculation, we shall also assume that the lines of the current are radial and that an allowance for the surface recombination gives only a small correction which does not distort greatly the excess hole distribution in the base. We shall introduce the concept of the volume recombination current, i_v, equal to the charge of holes recombining in the interior of the base per unit time, and we shall define similarly the surface recombination current i_R.

The value of i_v is given by

$$i_v = \frac{1}{\tau_p} Q_{st} = i_f . \tag{4.10}$$

The equation $i_v = i_f$ is self-evident since the whole forward current of the diode is carried by holes injected by the p-n junction into the base and all the holes introduced into the base recombine sooner or later.

The total surface recombination current is

$$i_R = \int\limits_{r_0}^{\infty} p(r)\, S_R 2\pi r\, dr = i_f\, \frac{S_R L_p}{D_p\,(1+A)}, \qquad (4.11)$$

where the distribution p(r), used in the calculation of the integral, is taken from Eq. (4.6). The role of the surface recombination is unimportant if $i_R \ll i_V$, i.e., when

$$S_R < \sqrt{\frac{D_p}{\tau_p}}\,(1+A). \qquad (4.12)$$

Using the maximum value of the hole lifetime in a point-contact diode, $\tau_p \approx 1\,\mu\text{sec}$, and assuming that $A \ll 1$, we find that the influence of the surface recombination can be neglected when $S_R \ll 7 \cdot 10^3$ cm/sec. Since practically all the etchants used in the preparation of point-contact diodes make the surface recombination velocity of germanium close to $S_R \approx 10^3$ cm/sec or less, it follows from this calculation that our assumption about the radial nature of the flow lines of the current is justified.

The relationship (4.12) is obtained for a steady-state forward current by estimating the influence of the surface recombination on the total value of the charge stored in the base. However, the response of a diode depends not only on the value of the stored charge but also on the charge distribution in the base. We can see from Fig. 4.2 that a considerable fraction of the charge in point-contact diodes is stored at distances from the rectifying contact such that the charge does not return to the p-n junction during switching but recombines in the base without making a contribution to the transient reverse current. Only that fraction of the charge which is stored in the immediate vicinity of the p-n junction affects the diode response; therefore, estimates obtained using Eq. (4.12) for steady-state conditions may be inaccurate or even completely wrong in the case of transient processes.

We shall assume arbitrarily that the diode response is affected only by the charge stored in a spherical layer extending radially from R = A to a distance ΔR. Integration of the expressions defining i_V and i_R, between the limits from R = A to R = A + ΔR, gives the following equation:

$$\frac{i_R}{i_v} = \frac{S_R L_p}{D_p} \frac{1 - e^{-\Delta R}}{1 - \left(1 + \frac{\Delta R}{1+A}\right) e^{-\Delta R}} \frac{1}{1+A}. \tag{4.13}$$

Assuming, as before, that the influence of the surface recombination can be neglected when $i_R < i_v$ in the volume of the base considered here, we obtain, instead of Eq. (4.12), a more exact condition:

$$S_R < \sqrt{\frac{D_p}{\tau_p}} (1+A) \frac{1 - \left(1 + \frac{\Delta R}{1+A}\right) e^{-\Delta R}}{1 - e^{-\Delta R}}. \tag{4.14}$$

If the radius of the p-n junction is not too small $[(r_0/L_p) \gtrsim 1]$, the transient processes are affected only by a layer whose thickness is close to the diffusion length of the holes, i.e., we may substitute $\Delta R = 1$ in Eq. (4.14). We can easily see that, in this case, the maximum values of S_R, estimated using Eqs. (4.14) and (4.12), differ by less than 30%, i.e., the difference is not very great.

When the radius of a rectifying contact is very small $[(r_0/L_p) \ll 1]$, the reduction in the hole density gradient near the p-n junction is mainly due to the diffusion whose effectiveness is governed by the characteristic length r_0. In this case, it is reasonable to assume that the charge in that part of the base which extends to three characteristic lengths takes part in the switching process, i.e., we may assume that $\Delta R = 3A$ in Eq. (4.14). Now, the estimates for the maximum value of S_R, obtained from Eqs. (4.12) and (4.14), respectively, differ by a factor of 7 for $A = 0.05$ and by a factor of 21 for $A = 0.01$. This means that, in order to neglect the influence of the surface recombination when $\tau_p = 1\,\mu\mathrm{sec}$ and the rectifying contact radius is very small ($A = 0.01$), we must satisfy the condition $S_R < 3 \cdot 10^2$ cm/sec. If, as before, the surface recombination velocity is $S_R = 10^3$ cm/sec, the hole recombination velocity in the active region of the base will be governed practically completely by the surface and not by the volume properties of the semiconductor.

Comparison of the conditions (4.12) and (4.14) shows that when the value of the radius of the p-n junction is small, it follows from Eq. (4.12) that the maximum permissible value of S_R is independent of r_0, but, according to Eq. (4.14), S_R depends very strongly on r_0: when r_0 is reduced the requirements that S_R must satisfy become more stringent if we still assume a spherical symmetry.

Physically, the increase in the importance of the surface recombination is explained by the fact that, when r_0 is reduced, the volume of the active part of the base decreases more rapidly (because it is proportional to r_0^3) than the surface (which is proportional to r_0^2).

In the light of this analysis, we must question the conclusion of Henderson and Tillman [30], drawn on the basis of Eq. (4.12), that the influence of the surface recombination on transient processes in point-contact germanium diodes can be neglected when the volume hole lifetime is $\tau_p < 40$ μsec.

In view of the considerable local overheating during the preparation of a p-n junction in a point-contact diode, the volume lifetime of holes τ_p is different in different parts of the base and its minimum value is observed in the contact region.

In view of the exceptional complexity of the recombination processes in the active part of the base, which is due to the appreciable importance of the surface recombination and the great variety of the values of the volume lifetime τ_p, it is usual to employ an effective lifetime in discussions of transient processes in point-contact diodes. For this reason, we shall assume τ_p to be the average effective lifetime representing the rate of hole recombination in the contact region.

§ 15. TRANSIENT CONDITIONS

15.1. Zero Load Resistance

We shall consider the switching of a diode with a hemispherical p-n junction from a steady forward state to the reverse direction in the circuit shown in Fig. 1.1 but we shall assume that $R_l = 0$. As in the case of a planar diode, this implies that the voltage across the p-n junction is reversed instantaneously, i.e., the boundary condition given by Eq. (4.5) is established instantaneously at the moment of switching. Thus, to find the function describing the time dependence of the reverse current, we have to solve the diffusion equation (4.2) with initial distribution given by Eq. (4.6) and the boundary condition given by Eq. (4.5).

The general method of solution consists of the transformation of Eq. (4.2) by the introduction of a new function

$$\varphi(R, \mathcal{J}) = p(R, \mathcal{J}) \cdot R. \tag{4.15}$$

After making some self-evident transformations, the problem can be be expressed in terms of the function $\varphi(R, \mathcal{J})$ in the following way:

$$\frac{\partial \varphi}{\partial \mathcal{J}} = \frac{\partial^2 \varphi}{\partial R^2} - \varphi, \tag{4.16}$$

$$\varphi(0, \mathcal{J}) = 0, \qquad \mathcal{J} \geqslant 0, \tag{4.17}$$

$$\varphi(R, 0) = \frac{j_f L_p}{q D_p} \frac{A^2}{1+A} e^{-(R-A)}, \quad \mathcal{J} = 0. \tag{4.18}$$

To within a constant factor in $\varphi(R, 0)$ and when R is measured from A, the expressions (4.16)–(4.18) are identical with Eq. (1.28), obeying the boundary conditions (1.30) and (1.31). Therefore, the solution for $\varphi(R, \mathcal{J})$ is found in the same way as the distribution of the hole density $p(X, \mathcal{J})$ in a planar diode with a semi-infinite base (cf. §3). After making the reverse substitution in accordance with Eq. (4.15), we obtain

$$p(R, \mathcal{J}) = \frac{p_1}{R} \psi_2 [(R-A), \mathcal{J}], \tag{4.19}$$

where $\psi_2[(R-A), \mathcal{J}]$ is defined by Eq. (1.51) and p_1 represents the impressed hole density near the p-n junction during the flow of the forward current; this density is found from Eq. (4.7).

The differentiation of $p(R, \mathcal{J})$ at R = A gives the following expression for the dependence of the transient reverse current density on time:

$$j(\mathcal{J}) = j_f \frac{A}{1+A} \left(\frac{e^{-\mathcal{J}}}{\sqrt{\pi \mathcal{J}}} - \text{erfc} \sqrt{\mathcal{J}} \right), \tag{4.20}$$

which is the same as the transient characteristic for a planar diode, given by Eq. (2.4), except that in the present case we have a factor

A/(1 + A) which is independent of time. When A →∞, the expression
(4.20) automatically reduces to (2.4); when $r_0 \ll L_p$, we find that
$A/(1 + A) \approx r_0/L_p$. Thus, the transient characteristic (4.20) is general and describes the decay of the reverse current in planar
(A → ∞) and point-contact (A ≪ 1) diodes. The time dependence of
the reverse current in point-contact diodes is exactly the same as
in planar diodes but the absolute values of j(t) are L_p/r_0 times
smaller in point-contact diodes.

It is evident from Eq. (4.20) that a diode may be regarded
as planar when $(r_0/L_p) > 2$-3, because in this case at any time after
switching the transient response differs from the response of a
true planar diode (A = ∞) by not more than 25-30%.

Asymptotic expressions for $j(\mathcal{J})$ in the case of small and
large values of $\mathcal{J}$ are:

$$j(t) \simeq j_f \ \frac{A}{1+A} \frac{e^{-\mathcal{J}}}{\sqrt{\pi\mathcal{J}}} \simeq \frac{r_0 j_f}{\sqrt{\pi D_p t}} e^{-t/\tau_p}, \qquad \mathcal{J} \ll 1, \qquad (4.21)$$

$$j(t) \simeq j_f \ \frac{A}{1+A} \frac{e^{-\mathcal{J}}}{2\sqrt{\pi\mathcal{J}^3}}, \qquad\qquad \mathcal{J} \gg 1. \qquad (4.22)$$

Since $\exp(-\mathcal{J}) \approx 1$, when $\mathcal{J} \ll 1$ it follows from Eq. (4.21) that,
during the initial phase after switching, the transient reverse current
is independent of the hole lifetime in the base but is governed solely by
the p-n junction radius. This can be explained physically as follows.
The time constants of the diffusion and recombination leveling of
the hole density near a p-n junction in point-contact diodes are
different and equal to, respectively, r_0^2/D_p and τ_p; for a thick-base
planar diode, both these time constants are identically equal to τ_p.
Therefore, in a point-contact diode during a time interval $t \ll \tau_p$,
the mechanism of the diffusion leveling of the hole density near the
p-n junction is the dominant process and this yields a dependence
j(t) of the type given by Eq. (4.21).

Thus, at a given time, the value of the reverse current should
be proportional to the forward current before switching and to the
radius of the rectifying contact.

Multiplying both parts of Eq. (4.20) by $\mathcal{J}^{3/2}$ and taking logarithms, we obtain the following expression:

$$\ln\left(\mathcal{J}^{3/2}\frac{i(t)}{i_f}\right) = \ln\frac{A}{1+A} + \ln\left[\mathcal{J}^{3/2}\left(\frac{e^{-\mathcal{J}}}{\sqrt{\pi\mathcal{J}}} - \text{erfc }\sqrt{\mathcal{J}}\right)\right]. \qquad (4.23)$$

We can easily show [30] that the function on the right-hand side of Eq. (4.23) has a pronounced maximum at $\mathcal{J} = 0.55$. Thus, if we plot an experimental dependence of the type $\ln[t^{3/2} i(t)/i_f] = f(t)$, its maximum should lie at $t = 0.55\tau_p$ irrespective of the radius of the p-n junction. This can be used to determine the hole lifetime in the base of a diode whose p-n junction has an unknown geometry.

We shall now find the recovered charge

$$Q_{\text{rec}} = \int_0^\infty i(t)\, dt = \frac{i_f\,\tau_p}{2}\frac{A}{1+A} = \frac{Q_{\text{st}}}{2}\frac{A}{1+A}. \qquad (4.24)$$

Thus, in spite of the fact that the stored charge is the same in planar and point-contact diodes (for the same values of the forward current and the same hole lifetimes in the base), the recovered charge, Q_{rec} transported by the transient reverse current after switching is L_p/r_0 times smaller in the case of a point-contact diode. Even when we take into account that, because of the high temperatures used during the electroforming of a point-contact diode, the hole lifetime in the base of such a diode may be considerably less than that in the base of a planar diode prepared from the semiconducting material, we can see why, during the early stages of the development of semiconductor electronics, only point-contact diodes were used in fast pulse circuits.

15.2. Switching After a Time Lag

We shall now consider the switching process after a certain time has elapsed between the end of a forward current pulse and the beginning of a reverse voltage pulse (cf. Fig. 2.5). By reducing Eq. (4.2) to Eq. (1.28), the hole distribution in the base of a point-contact diode during the delay and after the application of a reverse voltage can be found and analyzed in exactly the same manner as has been done in §4 for a planar diode. It is found [30] that in the tail end of the reverse current decay, when $\mathcal{J} \gg 1$ (we must stress here that $\mathcal{J}$ is measured from the end of the time lag, whose normalized duration is denoted by $\mathcal{J}_d$), we obtain — as in the case of a planar diode [cf. Eq. (2.24)]:

$$j(\mathcal{J}_d, \mathcal{J}) \simeq j_f \frac{e^{-\mathcal{J}}}{\sqrt{\pi \mathcal{J}^{3/2}}} K_2, \qquad (4.25)$$

where K_2 is a coefficient independent of time and governed only by the values of A and $\mathcal{J}_d$; for arbitrary time lags, this coefficient is given by

$$K_2(A, \mathcal{J}_d) = \frac{A}{1-A^2}\left[(1-A^2)e^{-\mathcal{J}_d} - A\,\mathrm{erfc}\,\sqrt{\mathcal{J}_d} + \right.$$

$$\left. + A^2 \exp\left(\mathcal{J}_d\,\frac{1-A^2}{A^2}\right) \mathrm{erfc}\,\frac{\sqrt{\mathcal{J}_d}}{A}\right]. \qquad (4.26)$$

We can easily see that when $A \to \infty$, $K_2(\mathcal{J}_d) = 1 - K_1(\mathcal{J}_d)$, where K_1 is defined by Eq. (2.14) and, therefore, Eq. (4.25) reduces to Eq. (2.24).

Immediately after switching, when the value of $\mathcal{J}$ is small, the form of the function $j(\mathcal{J}_d, \mathcal{J})$ can be obtained only by numerical integration of a general analytic expression describing the behavior of $j(\mathcal{J}_d, \mathcal{J})$ for arbitrary values of $\mathcal{J}$. The calculations carried out for two diodes [30] are shown in Fig. 4.3. It is interesting to compare the results given in that figure and in Fig. 2.6. When the delay time (time lag) is $t_d = \tau_p$, the reverse current after a time interval $t = 0.2\,\tau_p$ from the moment of switching is 0.15 i_f for a planar diode, 0.02 i_f for a diode with $r_0 = 0.5\,L_p$, and 0.002 i_f for a diode with $r_0 = 0.05\,L_p$. The relative fall of the current from the end of the time lag can be represented by factors of 3.5, 9, and 14, respectively, for these three diodes. It is evident from these examples that the lower the value of r_0/L_p, the more rapid is the re-establishment of the equilibrium distribution of holes in the diode during the time lag. In other words, even when switching is preceded by a time lag, the response time of point-contact diodes is less than that of planar diodes with the same value of τ_p and the time lag simply makes this difference greater.

15.3. Recovery Phase

When switching takes place in a circuit with a limiting resistance, it is interesting to determine the duration of the recovery phase when the transient reverse current is constant. As in the case of a planar diode, the hole distribution during this stage is described by the solution of Eq. (4.2) with the initial distribution given by Eq. (4.6) and the boundary condition at the p–n junction given by Eq. (4.3), where j is replaced with j_0. Such a solution has been obtained in [8, 30]; it is cumbersome and, therefore, we shall not give it here. Assuming that $p(r_0) = 0$ at $t = t_1$, we obtain an equation relating the duration of the first (recovery) phase $\mathcal{T}_1$ to the radius of the p–n junction and the switching parameter $(B = j_0/j_f)$:

$$\frac{1}{1+B} = \frac{A}{A-1}\,\mathrm{erf}\,\sqrt{\mathcal{T}_1} + \frac{1}{1-A}\left[1 - \exp\left(\mathcal{T}_1\,\frac{1-A^2}{A^2}\right)\mathrm{erfc}\,\frac{\sqrt{\mathcal{T}_1}}{A}\right]. \quad (4.27)$$

When $A \to \infty$, Eq. (4.27) reduces, as expected, to Eq. (2.28). The dependences of $\mathcal{T}_1$ on B for various values of A, plotted on the basis of Eq. (4.27), are shown in Fig. 2.8. Examination of these dependences indicates that when $0.1 < B < 1.0$ and $0.01 < A < 0.1$ (i.e., in the case of diodes with very small radii of the p–n junctions), the

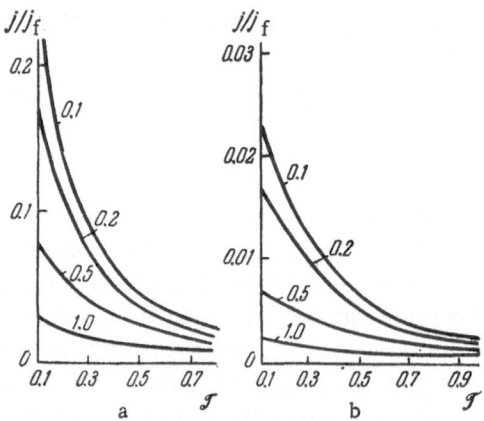

Fig. 4.3. Decay of the transient reverse current after various normalized delay times $\mathcal{T}_d$ (given alongside the curves) for two point-contact diode with A = 0.5 (a) and A = 0.05 (b).

where the approximate equality applies to $\mathcal{J} \gg 1$. It is interesting
to note that the approximate equality in Eq. (4.32) is identical with
Eq. (2.52), i.e., some time after the beginning of the transient process the postinjection emf decay curve of a point–contact diode has
a nearly linear part with a slope governed by the hole lifetime in
the base. In practice, the linear part of the postinjection voltage
decay in a point–contact diode cannot be observed at all because of
the immediate transition from the initial stage, described by Eq.
(4.30), to the final stage during which u(t) is again nonlinear because
$u(t) \leq kT/q$.

The transient characteristics u(t), calculated using the exact
expression (4.29), are shown in Fig. 4.4 for several diodes with
hemispherical junctions. Examination of the curves in Fig. 4.4
confirms that the relationships (4.30) and (4.32) can be used to describe approximately certain parts of the transient characteristics.

15.5. Establishment of a Steady-State Distribution

In our discussion of all the transient processes in point–contact diodes we have assumed that such diodes are switched from
the state during which a forward current has been flowing for an
infinitely long time, so that a steady-state hole distribution, of the
type given by Eq. (4.6), has been established in the base.

We shall now estimate the minimum duration of the forward
current pulse which produces a hole distribution in the base close

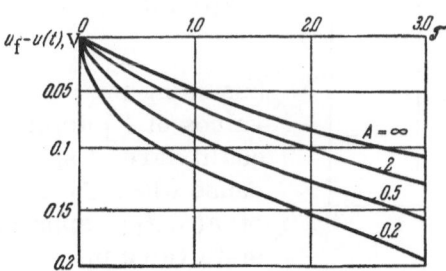

Fig. 4.4. Postinjection emf in diodes with
hemispherical p-n junction.

to a steady-state distribution. The solution of the diffusion equation (4.2) with zero initial distribution and the boundary condition of the type given by Eq. (4.3), where j is replaced with j_f, yields the following expression for the time dependence of the hole density near the p-n junction:

$$p(A, \mathcal{J}_f) = p_1 \left\{ \frac{A}{A-1} \operatorname{erf} \sqrt{\mathcal{J}_f} + \right.$$

$$\left. + \frac{1}{A-1} \left[\exp\left(\mathcal{J}_f \frac{1-A^2}{A^2}\right) \operatorname{erfc} \frac{\sqrt{\mathcal{J}_f}}{A} - 1 \right] \right\}, \qquad (4.33)$$

where p_1 has a definite value, which differs from diode to diode and which is given by Eq. (4.7). It is evident from Eq. (4.33) that initially $p(A) = 0$ and that $p(A) \to p_1$ when $\mathcal{J}_f \to \infty$. If we assume that $A = \infty$, we find that Eq. (4.33) reduces to Eq. (1.58), which describes the rise of the impressed hole density near the p-n junction when a forward current pulse is applied to a planar diode.

Figure (4.5) shows the time dependences $p(\mathcal{J})$ after the beginning of the flow of the forward current for various diodes; these dependences are plotted using Eq. (4.33). When the radius of the rectifying contact is reduced, the time taken to reach a steady-state distribution decreases. Comparison of the curves $p(A, \mathcal{J}_f)$ for large values of A shows that, in practice, we may assume that diodes with A ≥ 2-3 are planar.

For a diode with a very small radius of the p-n junction (A ≪ 1), Eq. (4.33) can be transformed to

$$p(A, \mathcal{J}_f) \simeq p_1 \left(1 - \frac{3A}{\sqrt{\pi \mathcal{J}_f}}\right), \qquad (4.34)$$

assuming that $A^2 < \mathcal{J}_f < 1$. The values of $\mathcal{J}_f$ defined by this inequality are of special interest because when $\mathcal{J}_f > A^2$, the values of $p(A, \mathcal{J}_f)$ approach the steady-state value p_1.

Assuming approximately that the value of the impressed hole

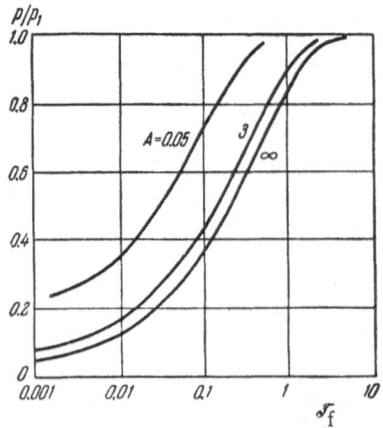

Fig. 4.5. Increase of the impressed hole density near the p-n junction during the flow of the forward current.

density near the p-n junction is close to its maximum value when $p(A) = 0.7p_1$, we find that this state is reached when the duration of the forward current pulse exceeds a minimum value defined by

$$t_f^{\min} \simeq 3A^2 \tau_p = 3\frac{r_0^2}{D_p}. \tag{4.35}$$

Thus, the time necessary for the establishment of a steady-state distribution of holes in the base of a point-contact diode during a forward current pulse is independent of the hole lifetime in the base and is governed only by the radius of the p-n junction or, more exactly, by the time constant r_0^2/D_p. Noting that for a planar diode $t_f^{(\min)} \approx 0.5\tau_p$, we find that the time for the establishment of a steady-state hole distribution in a point-contact diode is $0.15/A^2$ times shorter than for a planar diode with the same value of the hole lifetime.

In all point-contact diodes used in practice, we have $r_0 < 25\mu$ and, therefore, $t_f^{(\min)}$ does not exceed several tenths of a microsecond so that in the great majority of cases the steady-state forward current conditions are achieved easily in point-contact diodes.

15.6. General Comparison with Planar Diodes

We shall now compare generally the transient processes in planar diodes with those in diodes having hemispherical contacts.

It follows from the foregoing discussion that the formulas for a particular transient process in a planar diode are obtained from a more general expression which is valid for arbirary values of r_0/L_p and formulas for the planar case are obtained by assuming that $(r_0/L_p) \to \infty$, i.e., under all conditions, a planar diode is a special case of a diode with a hemispherical contact of arbitrary radius. In practice, we may assume that a diode is planar when $A \geq 2\text{-}3$, because under these conditions we can deduce the various parameters representing transient processes with an accuracy of 20-30%. These parameters may be the values of Q_{rec}, t_1, $i(t)$ at a given moment, $\partial u(t)/\partial t$, etc. If an accuracy to within a factor of 2 is acceptable, we can use the formulas for the planar diode even when $A \geq 1$.

On the other hand, when A < 0.5, the durations of various transient processes are considerably shorter than those in the case A = ∞ and the difference between point-contact and planar diodes becomes noticeable in experimental investigations.

The values of the transient reverse current under any switching conditions is proportional to the forward current in planar and in contact diodes: all the formulas include only the ratios $j(t)/j_f = i(t)/i_f$ and not the two values of the current separately.

In the initial stage of any transient process in a point-contact diode, the rate of establishment of a new equilibrium state is determined by the time constant r_0^2/D_p, which is related to the contact geometry but is independent of the hole lifetime in the base. The duration of the constant-reverse-current (recovery) phase t_1, the decay time of the postinjection emf, the time taken to establish a steady-state hole density near the p-n junction during a forward current pulse, are all directly proportional to the time constant r_0^2/D_p; the values of the reverse current immediately after switching (in the case $R_l = 0$) are proportional to $\sqrt{r_0^2/D_p}$.

After a sufficient time interval from the moment of switching, the time dependences of various characteristics begin to be governed by the relaxation time constant τ_p, i.e., they obey the same relationships as the characteristics of a planar diode with the same hole lifetime in the base. The only difference is that the absolute values of the transient currents or voltages in the case of a point-contact diode at a given value of t are always smaller than those for an analogous planar diode and this difference increases with the decrease of the value of r_0/L_p. It is interesting to note that the recovered charge, which is an integral characteristic of the transient switching process, depends in the same way on both time constants r_0^2/D_p and τ_p. In fact, when A ≪ 1, we find that Eq. (4.24) yields

$$Q_{rec} \simeq \frac{i_f \tau_p}{2} A = \frac{i_f}{2} \sqrt{\tau_p} \sqrt{\frac{r_0^2}{D_p}}. \tag{4.36}$$

These properties of point-contact diodes are due to the fact that during the initial stage of any transient process the establishment of an equilibrium distribution of holes near a p-n junction is governed

primarily by the diffusion leveling of the hole density whose effectiveness increases with decrease of the time constant r_0^2/D_p.

§ 16. EXPERIMENTAL INVESTIGATIONS

Experimental investigations of the transient processes in
point-contact diodes have been carried out by many investigators,
beginning from the late forties. However, not all the results obtained can be compared with the theory because of the extreme
complexity of the investigated devices (compared with the simple theoretical models). Consequently, many important diode parameters
are indeterminate.

16.1. Observations of the Transient
Reverse Current Decay

Waltz [95] observed transient processes in electroformed
point-contact diodes prepared from n-type germanium with ρ =
10 Ω·cm. The diameter of the formed p-type region was close to
50 μ but the hole lifetime was not measured. The reverse current
decay curve of all diodes was found to consist of two regions: an
initial very large peak, ending approximately in 0.01 μsec, followed
by a "tail" lasting at least 0.05 μsec. Waltz concluded that his
point-contact diodes had a very low injection efficiency ($\gamma \approx 0.02$)
and, therefore, that electrons were accumulated in the p-type region during the flow of the forward current. In his opinion, this
accounted for the initial short but sharp peak of the forward current. Holes in the diode base were also accumulated during the
flow of the forward current, mainly at various traps generated in
germanium by the electroforming process. The liberation of holes
from the traps with a time constant of ≈ 0.05 μsec accounted for
the "tail" in the transient reverse current characteristic.

A different explanation of the experimental results of Waltz
was given by Henderson and Tillman [30]. They criticized the trap
mechanism of the accumulation of holes in the base. When holes
are accumulated at traps, the charge should be independent of forward current i_f when this current is sufficiently large to saturate
all traps; in practice, the charge is almost directly proportional
to $i(t)$ and i_f during the initial and final stage of the transient. It
is more reasonable to assume that the injection efficiency of the
investigated electroformed contacts is smaller than 1 but much

greater than 0.02. In this case, the decay of the reverse current in the tail part of the transient process can be explained by the accumulation and dispersal of holes in the base in accordance with Eqs. (4.20) or (4.22). To explain Waltz's results quantitatively, it is necessary to assume that $\tau_p = 0.3\ \mu\text{sec}$, which is quite reasonable for some fast-response diodes. The direct proportionality of the transient reverse current to the forward bias was reported also in [56]. Measurements were carried out on commercial point-contact germanium diodes in the forward current range $i_f = 5\text{-}30$ mA using apparatus with a resolution time shorter than 0.02 μsec.

Special experiments, whose purpose was to check the theoretical formulas, were carried out by Armstrong, Metz, and Weiman [102]. Figure 4.6 shows the dependence of $\ln[i(t)\sqrt{t}]$ on t obtained by these authors for a particular point-contact diode. The linearity of this dependence confirms the validity of Eq. (4.21) describing the initial stage of the transient process. The effective hole lifetime in the base, found from the slope of the straight line in Fig. 4.6, is $\tau_p = 2\ \mu\text{sec}$, which is in agreement with the results of other experiments.

Armstrong et al. [102] used, as one of the parameters of the transient process, the reverse current i_g measured 2 μsec after switching of the diode by a reverse voltage. A calculation using Eq. (4.21) showed that the value of i_g should be 3.6 mA for $i_f = 30$ mA and $r_0 = 25\ \mu$. The experimental value was $i_g \approx 1.2$ mA.

The reason for the difference between the calculated and experimental values was not explained in [102]. Actually, the reason

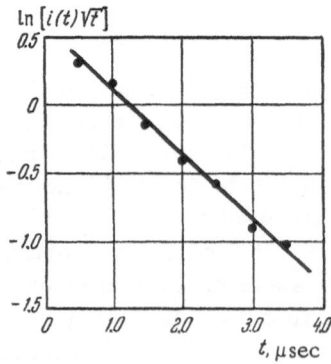

Fig. 4.6. Time dependence of the transient reverse current in a point-contact diode with $\tau_p = 2\ \mu\text{sec}$.

is trivial and is due to the fact that the exact formula (4.20), valid
for any value of A and $\mathcal{J}$, should have been used instead of Eq.
(4.21). If we apply the exact formula to $r_0 = 25\,\mu$, $\tau_p = 2\,\mu$sec, and
$i_f = 30$ mA and use curve 1 in Fig. 2.1, we obtain $i_{\mathscr{G}} \approx 1$ mA, which
is practically equal to the experimentally obtained value of this parameter.

The preparation and investigation of special batches of point-
contact diodes with electroformed contacts of 25–50 μ diameters
showed that the quatities $i_{\mathscr{G}}$ and r_0 were directly proportional, as
predicted by Eq. (4.21).

Thus, the experiments of Armstrong et al. [102] confirmed
the validity of the dependence of the transient reverse current on
r_0, t, τ_p, and i_f, given by Eq. (4.20).

Some investigations of commercial point-contact germanium
diodes were carried out by Henderson and Tillman [30] using a
sensitized stroboscopic method (resolution time less than $0.015\,\mu$sec).
They found qualitative agreement with the theory but the quantitative discrepancies were quite considerable. In order to obtain
agreement between the theory and the experimental results (obtained
for diodes prepared under identical conditions), it was necessary
to assume that the parameter τ_p could differ by up to a factor of
10 in these diodes.

The experiments described show that it is very difficult to
check the theoretical formulas using the results obtained for commercial point-contact diodes since such important characteristics
as τ_p and r_0 are not known and the value of the injection efficiency
γ is also effectively unknown. Without the knowledge of the values
of these quantities, very arbitrary theoretical interpretations are
possible of the obtained experimental results.

16.2. Investigations of Microalloyed Diodes

Henderson and Tillman [30] checked Eq. (4.20) using specially
prepared diodes with hemispherical rectifying contacts. Diodes
with small p-n junction areas were prepared by alloying indium to
n-type germanium in which the hole lifetime was $\tau_p = 500\,\mu$sec.
Alloying reduced strongly the value of τ_p but this value remained
at a level of 11 $\pm$ 2 μsec, as established by measurements of the

diffusion length by the light spot method. The indium-alloyed regions were nearly hemispherical and their radii were 0.1, 0.2, and 0.29 mm, respectively, for the three diodes investigated. Thus, the values of A for these samples were within the limits 0.5-1.5. After etching, all the p-n junctions exhibited current-voltage characteristics with a pronounced reverse-current saturation region, which indicated that the influence of surface leakage was negligibly small.

Henderson and Tillman recorded the time dependences of the transient reverse current after the switching from a forward current $i_f \approx 10$ mA by a reverse voltage of 7 V. The resistance in the diode circuit was so small that the duration of the constant-reverse-current (recovery) phase was less than $t_1 = 0.1\,\mu$sec, i.e., in analysis of processes at $t \gg 0.1\,\mu$sec we could assume that the switching took place in a circuit with $R_l = 0$. Henderson and Tillman were unable to determine the i(t) curve after a time lag between the end of the forward and reverse pulses because their apparatus was insufficiently sensitive although they used the stroboscopic method.

The experimental results were used to plot the dependences $\ln\,[i(t)t^{3/2}/i_f]$ on t, shown in Fig. 4.7. The values of τ_p, determined from the maxima of these curves [cf. Eq. (4.23) and the discussion which follows it], were 11, 9, and 12 μsec for the three investigated diodes; these values were not in satisfactory agreement with $\tau_p = 11 \pm 2\,\mu$sec found by the light spot method. The values of τ_p found by analysis of the "tail" parts of the i(t) curves at large values of t were somewhat smaller: 7, 7.5, and 6.5 μsec for the same diodes. The cause of this difference may have been the low accuracy of the determination of the function i(t) at high values of t,

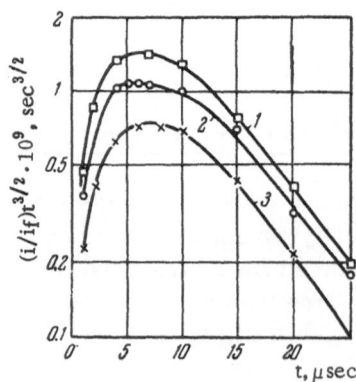

Fig. 4.7. Dependence of $(i/i_f)t^{3/2}$ on t for diodes with different values of r_0 (mm): 1) 0.29; 2) 0.20; 3) 0.10.

when the useful signal was comparable with the intrinsic noise of the measuring device.

Examination of the curves plotted in Fig. 4.7 confirms exact quantitative agreement between the measured values of i(t) and the theoretical formulas. According to Eq. (4.20), the ratio of the values of i(t) at any fixed value of t should be 1:1.5:1.8 for the three diodes (A = 0.5, 1.0, and 1.5). The experimental curves at t = 10 μsec show that this ratio is 1:1.45:1.85. The experimental value of $i(t)/i_f$ = 0.5 at t = 1 μsec for the diode with A = 1.5 is in good agreement with 0.6, calculated from Eq. (4.20).

The switching of the same microalloyed diodes was investigated in a circuit with a large load resistance and under open-circuit conditions when the forward current was switched off. Dependences of the duration of the constant-reverse-current (recovery) phase on the switching conditions are shown for all three diodes in Fig. 4.8. Analysis of the dependences in this figure gives τ_p = 12, 9, and 10 μsec, respectively; these values are close to the lifetimes obtained under switching conditions with R_l = 0.

Observation, under open-circuit conditions, of the postinjection emf decay after the end of a forward current pulse confirmed, at least qualitatively, the principal theoretical predictions. After the initial sudden change of the voltage across the diode, associated with the ohmic voltage drop across the base resistance, a rapid nonlinear decay u(t) was observed: the smaller the radius of the alloyed contact, the faster was this decay stage. Later, at t > 10–15 μsec,

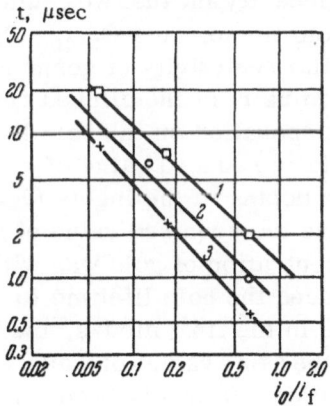

Fig. 4.8. Dependence of the duration of the first (recovery) phase on i_0/i_f for three microalloyed diodes with different values of r_0 (mm): 1) 0.29: 2) 0.20: 3) 0.10.

the u(t) curves became linear and their slopes indicated values of $\tau_p = 9.5$, 7, and 8 μsec. The experimental points fitted well the theoretical curves calculated using Eq. (4.29) and the values of τ_p which have been just quoted. The agreement between the calculated and measured values was observed in the linear and nonlinear stages of the decay right up to $\tau = 3\tau_p$.

Summarizing the results of all these experiments, we may conclude that the hole lifetimes found by analysis of the transient reponse obtained under various swithching conditions are 9.5-12 μsec for a diode with r = 0.029 cm, 8-12 μsec for $r_0 = 0.020$ cm, and 7-9 μsec for $r_0 = 0.010$ cm. Thus, the values of τ_p obtained under different conditions agreed to within ± 10%.

Studies of the transient processes in diodes with hemispherical contacts have been carried out also by the present author and Postnikova. Our measurements were carried out on germanium diodes with gold-wire contacts. A rectifying contact was obtained by welding a wire to a germanium crystal by heating the point of contact with an electric current pulse. The wire was made of an alloy of gold and gallium, which ensured a high conductivity of the recrystallized p-type region and a high injection efficiency $\gamma = 1$ for all the forward currents passed through the diodes during measurements.

The rectifying-contact radius was determined by measuring the dependence of the diode capacitance C_d on the reverse voltage u_r. By plotting this dependence in the coordinates $1/C_d^2$ and u_r, we could determine the capacitance of the diode encapsulation C_e from the deviation of the dependence from linearity and then we could find the p-n junction capacitance, C_{p-n}, from the formula $C_{p-n} = C_d - C_e$. The capacitance C_{p-n} and the initial resistivity of germanium were used to determine the average radius r_0 of the alloyed region. To reduce the influence of the welding operation on the hole lifetime, we used germanium with a specially reduced value of τ_p. This reduction was achieved by double doping of the ingots (during their pulling) with antimony, which gave the required value of ρ, and gold, which reduced τ_p. The concentration of gold was within the limits $(2-5) \cdot 10^{14}$ cm^{-3} and it reduced the hole lifetime to $\tau_p \approx 0.0.5-0.2$ μsec. The hole lifetime in the final diodes, made from the same germanium ingot with the same values of ρ and N_{Au}, varied only slightly from diode to diode (this was specially checked).

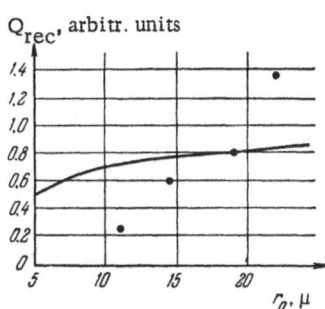

Fig. 4.9. Dependence of the recovered charge on the radii of hemispherical p-n junctions (the
theoretical curve is shown continuous and the points represent experimental values).

We used the recovered charge Q_{rec} as the parameter representing the diode response. The value of Q_{rec} was measured by
switching from $i_f = 2$ mA to $U_r = 10$ V; the error in the measurement of Q_{rec} did not exceed 5 nC. When the initial hole lifetime in
germanium was $\tau_p \approx 0.1$ μsec, we found that the experimental points
agreed satisfactorily with the theoretical curve plotted on the basis
of Eq. (4.24).

When the hole lifetime τ_p was shorter ($\sim 5 \cdot 10^{-9}$ sec), the dependence of Q_{rec} on r_0 was stronger (Fig. 4.9) than the theoretically
predicted dependence. We were unable satisfactorily to explain
this discrepancy.

The experiments reviewed in the present section show that in
all known types of diode with hemispherical p-n junctions (point-
contact diodes with electroformed contacts, microalloyed diodes,
welded-contact diodes), the transient processes were described
satisfactorily (at least semiquantitatively) by the the theoretical
formulas given in § 15.

Chapter V

Effect of an Electric Field in a
Diode Base on Transient Processes

§ 17. BUILT-IN INTERNAL FIELD
IN A DIODE BASE

In the preceding chapters, we have assumed that the base of a diode is uniform in its electrical properties and, in particular, that the distribution of the donor impurities as well as the resistivity ρ_n are uniform.

There are, however, many devices which do not satisfy this assumption. In particular, they include diffused diodes and transistors.

All types of nonuniform distribution of the donor impurity concentration in the base can be divided into two categories: an increasing value for N_d away from the p–n junction in the direction of the body of the base; and a decreasing value for N_d along the same direction. The first category includes those diodes in which the p–n junctions are prepared by the diffusion of an acceptor impurity in a uniform n-type semiconductor (diffused diodes, collector p–n junctions of drift transisitors). In devices of the second category, the base region itself is prepared by diffusion and the heavily doped p-type region is prepared either by additional diffusion or by alloying (emitter p–n junctions of drift transistors). The two different forms of the distribution of the active impurity concentration in diodes are shown in Fig. 5.1.

It is known that an electric field always exists in the body of a semiconductor with a nonuniform distribution of donors or

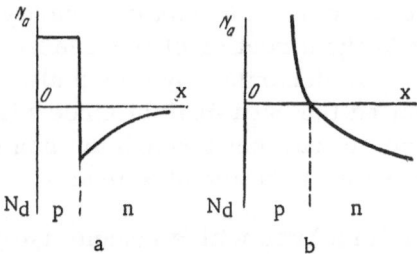

Fig. 5.1. Nonuniform distribution of a donor impurity in diffused diodes with accelerating (a) and retarding (b) fields.

acceptor impurities. Let us consider an n-type semiconductor whose properties are far removed from intrinsic conduction and in which the donor impurity concentration N_d varies monotonically along the x axis.

In accordance with the principle of neutrality, the following relationship should be satisfied at each point in the base:

$$n_{n0}(x) \simeq N_d(x). \tag{5.1}$$

Under thermodynamic equilibrium conditions a diffusion flux of free electrons along the direction of decreasing electron density is balanced by a drift flux of electrons in the opposite direction, so that the total electron current is equal to zero. Equation (1.4) for the equilibrium state (i.e., in the absence of a current) is

$$q\mu_n n_{n0}(x)E + qD_n \frac{\partial n}{\partial x} = 0. \tag{5.2}$$

Using Eq. (5.1) and the Einstein relationship $[D_n = (kT/q)\,\mu_n]$, we obtain the following expression for the electric field intensity in the base:

$$E(x) = -\frac{kT}{q}\frac{1}{N_d(x)}\frac{dN_d(x)}{dx}. \tag{5.3}$$

Thus, the presence of a donor concentration gradient in an n-type semiconductor produces an electric field pointing along the direction of the decreasing impurity concentration. We can easily show that in a p-type semiconductor an acceptor impurity gradient produces a field pointing in the direction of increasing impurity concentration and this field is described by a formula similar to Eq. (5.3). This field is called the built-in field since it is not associated with the flow of an electric current through the semiconductor but is established during the preparation of a diode.

An electric field in a base with a nonuniformly distributed donor impurity alters the equilibrium distribution and the rate of establishment of the equilibrium of the excess holes injected by the p-n junction. An electric field in a nonuniformly doped semiconductor opposes the flow of minority carriers directed along the increasing impurity concentration, and aids the flow of minority carriers directed along the decreasing impurity concentration. Depending on the the effect of the electric field in the diode base on the flow of minority carriers, the built-in field is called retarding or accelerating.

Equation (5.3) remains valid for all those parts of the base where $n_{n_0}(x) \gg p(x)$, i.e., when the injection level is low. When the injection level is increased so that the density of excess holes becomes comparable with or greater than the concentration of donor atoms, the field produced by a nonuniform distribution of donors becomes weaker and eventually ceases to have any effect altogether.

Therefore, all processes involving a built-in field in the base are only of interest under low injection conditions.

The nature of the effect of the built-in field on transient processes in a diode is governed by its intensity and particularly by its direction.

When a forward current flows through a diode with a retarding field in its base, this field opposes the diffusion flow of injected holes from the p-n junction into the body of the base. Excess holes are "pushed" by the field toward the junction; thus, the hole density near the junction is increased and the hole density in the body of the base is reduced compared with the case when there is not such field in the base. The degree of change in the hole distribution depends on the intensity of the retarding field.

Changes in the nature of the excess hole distribution during the flow of a forward current produce corresponding changes in the transient process when a diode is switched by a reverse voltage.

The higher hole density near the p-n junction increases the time necessary for the extraction of holes (during the first phase) which reduces to zero the hole density at the p-n junction. Holes in the peripheral parts of the base drift to the p-n junction under the action of the retarding field during the first phase of the transient process. This also increases the duration of the first stage.

Such an increase in the duration of the first switching phase (during which the reverse conductance is high) means that the fraction of excess holes extracted during this phase is considerably greater than in a diode without a retarding field in the base. Thus, the presence of a retarding field reduces the number of excess holes remaining in the base at the end of the first phase of the switching transient. Therefore, the duration of the second phase (during which the reverse resistance of the diode is restored) is reduced by a retarding field.

The drift flow of holes to the p-n junction reduces the importance of recombination, and this increases the recovered charge compared with the case of a field-free base.

In some respects a diode with a retarding field in the base is similar to a thin-base diode with a noninjecting ohmic contact (Chap. III). In both devices, the injected excess holes are localized (by different means) in the direct vicinity of the p-n junction.

When a forward current flows through a diode, an accelerating field in the base tends to disperse more rapidly the holes injected by the p-n junction. Therefore, the steady-state impressed density of holes near the p-n junction is lower than in a diode with a field-free base, but in the peripheral regions of the base, the density of holes is correspondingly higher.

This discussion shows clearly that during the switching by a reverse voltage the duration of the first (recovery) phase will be shorter than in a diode with a retarding field.

We can say nothing definite about the duration of the second phase in the presence of an accelerating field in the base; in fact, special calculations are required because, on the one hand, the total number of holes in the base at the beginning of this phase is higher than in a field-free diode, but, on the other hand, the fraction of holes reaching the p-n junction is less because holes are located further away from the p-n junction and they tend to drift away from the junction.

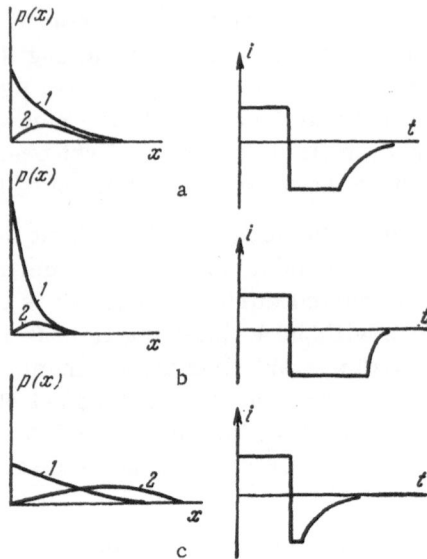

Fig. 5.2. Distribution of holes in the base during the flow of a forward current (1) and at the end of the first (recovery) phase (2), with corresponding switching oscillograms of diodes with a uniform base (a), a base with a retarding field (b), and a base with an accelerating field (c).

The motion of holes away from the p-n junction throughout the whole transient process increases the importance of recombination and reduces the value of the recovered charge compared with the case of a diode having a uniform base.

We may assume that, from the point of view of a switching transient, a diode with an accelerating field in the base is similar to a diode with a small-area hemispherical contact. In both devices, the accumulation of the excess hole charge takes place in the peripheral regions of the base and, moreover, holes are pulled away from the p-n junction much more strongly than by the diffusion forces in a planar diode with a uniform base.

The difference between the nature of the distribution of the excess holes and the changes in the transient reverse current (during switching in a circuit shown in Fig. 1.1) are shown in Fig. 5.2 for the three categories of diodes considered here.

§ 18. FORWARD-BIASED DIODE WITH A BUILT-IN FIELD

We shall now find the steady-state distribution of injected holes in the base of a diode with a built-in field, biased in the forward direction. We shall use a planar diode model with a semi-infinite base, which has been discussed in § 3.

In addition to the assumptions made earlier, we shall consider also the effects of the presence of an electric field in the base. For simplicity, we shall assume that there is a constant electric field E in the base; we shall also assume that this field is directed at right-hand angles to the p-n junction plane and that it is not affected by the injection of holes. When E > 0, the field is accelerating; but when E < 0, the field is retarding.

It follows from Eq. (5.3) that the electric field intensity is independent of the position coordinate if the donor impurity distribution in the base obeys an exponential law:

$$N_{\mathrm{d}}(x) = N_{\mathrm{d}}(0)\, e^{\varkappa x}. \tag{5.4}$$

A change in the impurity concentration, introduced into a semiconductor by the diffusion method, is simply described by the complementary error function, which can often be replaced by an exponential function without committing a serious error. Therefore, we may assume that in real diffused diodes, the field in the base is more or less constant.

When these assumptions are made, we find that the behavior of holes in the diode base is described by the diffusion-drift equation (1.11). The second term on the right-hand side of this equation allows for the influence of the electric field on the hole density distribution. The boundary condition in the p-n junction plane during the flow of a forward current of density j_{f} is of the form

$$-q D_{p} \frac{\partial p}{\partial x} + q \mu_{p} E p = j_{\mathrm{f}} \qquad \text{when} \quad x = 0, \tag{5.5}$$

but when a specified voltage is applied to the p-n junction, the boundary condition is in the form given by Eq. (1.20).

The solution of Eq. (1.11) in the steady-state case ($\partial p/\partial t = 0$) with the zero initial distribution and the boundary condition given by Eq. (5.5) is

$$p(X) = p_1\left(\sqrt{1+E_n^2} - E_n\right)\exp\left[X\left(E_n - \sqrt{1+E_n^2}\right)\right], \qquad (5.6)$$

where, as before, $X = x/L_p$ and p_1 is given by Eq. (1.26). The quantity E_n represents the normalized electric field, defined by

$$E_n = \frac{qL_p}{2kT}E. \qquad (5.7)$$

The solution (5.6) was obtained first by Moll, Krakauer and Shen [105]. Thus, the distribution of excess holes in the base of a diode with a built-in field is described by an exponential function, i.e., this distribution is of the same nature as that in a diode with a uniform base. The pre-exponential term, i.e., the density of holes near the p-n junction, is greater than p_1 in the presence of a retarding field and smaller than p_1 in the presence of an accelerating field. The index of the exponential function is negative, irrespective of the field direction, so that the density of holes decreases away from the p-n junction. The characteristic length of this decrease is greater in the presence of an accelerating field and smaller in the presence of a retarding field.

When the built-in field is large, $|E_n| \gg 1$, Eq. (5.6) can be simplified:

$$p(X) = 2p_1|E_n|\exp(-2X|E_n|) \quad \text{when} \quad E_n < 0, \qquad (5.8)$$

$$p(X) = \frac{p_1}{2E_n}\exp\left(-\frac{X}{2E_n}\right) \qquad \text{when} \quad E_n > 0. \qquad (5.9)$$

Thus, in a strong built-in field the impressed density of holes near the p-n junction increases or decreases by a factor of $2E_n$ and the "effective" diffusion length decreases or increases by the same factor of $2E_n$ in the presence of a retarding or accelerating field. The "effective" diffusion length is the distance at which the density of injected carriers decreases by a factor of e from its initial value. If the built-in field is weak, i.e., $|E_n| < 1$, it has practically no effect on the hole distribution so that this distribution is governed solely by the diffusion length of holes and described by the relationship (1.24).

It is evident from Eq. (5.7) that the concepts of "strong" and "weak" fields are relative and are governed not only by the absolute value of the built-in field intensity but also by the diffusion length of holes in the base. Physically, this is self-evident because either recombination or drift processes dominate the distribution of holes, depending which of the characteristic lengths L_p or kT/qE is the smaller.

We shall now estimate the rate of establishment of a steady-state hole distribution in the base of a diode with a built-in field after the application of a forward bias. A mathematical solution of this problem was obtained by Gaman [106] for a more complex model of a diode than that employed by us. Gaman considered a planar diode with a base of arbitrary thickness W and assumed an arbitrary surface recombination velocity S_R in the plane of the ohmic contact. The other assumptions made by Gaman were identical with those stated at the beginning of the present section.

Gaman obtained general expressions for p(x, t) in the case of a sudden change of a forward voltage across the p-n junction, as well as a sudden change of the forward current through a diode from zero to some specified value. Gaman's expressions are very cumbersome and they will not be given here.

Analysis of Gaman's expressions shows that the time dependences of the hole density gradient near the p-n junction and, therefore, the time dependences of the forward current (in the case of the application of a voltage step) and the impressed hole density near the p-n junction as well as the time dependences of the forward voltage across the p-n junction (in the case of a current step) are governed by the time constant

$$\tau_{WE0} = \left(\frac{\lambda_m^2 D_p}{W^2} + \frac{(\mu_p E)^2}{4 D_p} + \frac{1}{\tau_p} \right)^{-1} \qquad (5.10)$$

at values of time which are not too close to the moment of application of the forward bias.

The quantity λ_m in Eq. (5.10) is the root of the transcendental
equation

$$\tan \lambda_m = \frac{\lambda_m}{\dfrac{qEW}{2kT} - \dfrac{S_R W}{D_p}}. \qquad (5.11)$$

Thus, in the presence of a built-in field, the transient processes
occurring during the switching of a diode from the neutral to the
forward state are accelerated if

$$\frac{(\mu_p E)^2}{4D_p} > \frac{\lambda_1^2 D_p}{W^2} + \frac{1}{\tau_p}, \qquad (5.12)$$

irrespective of the direction of the built-in field.

In the case of a diode with a semi-infinite base ($W \to \infty$), the
strong-field condition, given by Eq. (5.12), reduces to its usual
form $E_n^2 > 1$ or $|E_n| > 1$.

Let us now analyze the process of accumulation of the excess
hole charge in the base.

Integration of Eq. (5.6) over the whole volume of the base
shows that the stored charge is independent of the sign and intensity of the built-in field and, as in the case of a diode with a uniform base, this charge is $Q_{st} = i_f \tau_p$. However, the distribution of
the stored charge depends considerably on the built-in field direc-
tion. Thus, when this field is not very strong, $|E_n| = 2$, more than
$0.95 Q_{st}$ is concentrated at a distance L_p from the p-n junction when
the field is retarding, but less than $0.2\, Q_{st}$ is present in the same
region when the field is accelerating.

It is interesting to note that the following very different diodes
can store the same amount of charge: a planar diode with a semi-
infinite base irrespective of whether there is a built-in field in the
base and whether this field is accelerating or retarding; a diode
with a hemispherical contact or with a semi-infinite base; a planar diode with a thin base and a noninjecting ohmic contact. When

the hole lifetime in the base and the forward current are the same for all these diodes, they store the same amount of charge $Q_{st} = i_f \tau_p$, as given by Eq. (1.6).

The stored charge decreases only when additional recombination centers (such as those found in the presence of a recombination-type ohmic contact in a thin-base diode) are present in the immediate vicinity of the p-n junction.

The value of the charge stored in a diode with a p-n junction and base of arbitrary geometry and with an arbitrary distribution of the electric field in the base is $Q_{st} = i_f \tau_p$ if additional recombination centers (i.e., additional to those originally present in a semiconductor) are not generated near the p-n junction.

This conclusion is valid for low injection levels and a diode in which τ_p has the same value throughout the base.*

On the other hand, the distribution of the charge stored in the base depends on the diode structure. Therefore, the recovered charge as well as the response of a diode are different for different structures.

The influence of a built-in accelerating field on the charge stored in diodes with a special distribution of ionized atoms, obtained by back-diffusion of an impurity from a semiconductor, was first investigated by Halpern and Rediker [83].

Later, Kennedy [107] carried out detailed numerical calculations, using an electronic computer, for diodes with an arbitrary base thickness and a built-in field. The results of Kennedy's calculations for a steady-state distribution are shown in Fig. 5.3; the curves in this figure give the dependences of the stored charge on the relative base thickness, the surface recombination velocity,

---

* For the most general diode model and an arbitrary forward current, the following self-evident observation can be made about the stored charge (this observation follows from analysis of the equation of continuity): during the prolonged flow of a forward current through a diode, the hole charge stored in the base is equal to the product of the average hole lifetime and the hole component of the forward current from which the total hole current flowing out through the boundaries of the semiconductor is subtracted. If these boundaries are sufficiently far from the p-n junction, the value of Q_{st} is given by the very simple expression (2.8) in which τ_p must be replaced with its average value.

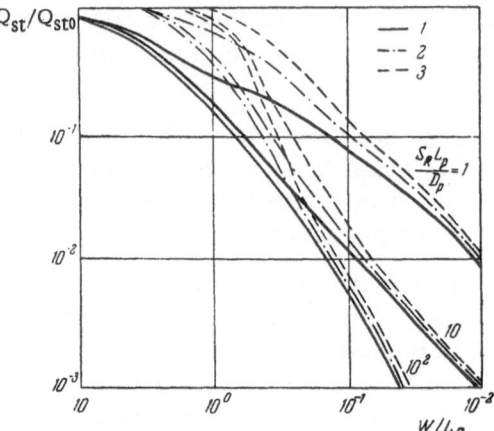

Fig. 5.3. Dependence of the stored charge on the base thickness for various values of the parameters $S_R L_p / D_p$ (given alongside the curves) and the following values of E_n: 1) +2.5; 2) 0; 3) -2.5.

and the built-in electric field intensity. The quantity Q_{st_0} represents the charge stored in a diode with an infinitely thick base.

Analysis of the dependences shown in Fig. 5.3 yields the following conclusions. In the presence of an accelerating field, an ohmic contact affects the storage of holes at base thicknesses greater than those in the case of diodes with uniform bases. A retarding field has an opposite effect. If we define the base to be "semi-infinite", i.e., of thickness equivalent to W_∞, when $Q_{st} = 0.7 Q_{st_0}$, we find that diodes with $E_n = +2.5$, 0, and -2.5 have, respectively, $W_\infty / L_p = 5$, 2, 0.5.

When the base thickness is reduced, the role of the built-in field E in the charge storage becomes less important irrespective of the sign of this field and of the value of the surface recombination velocity in the ohmic contact plane. For sufficiently small values of W [$(W/L_p) \lesssim 0.1$ in Fig. 5.3], the stored charge is governed only by the parameters W/L_p and $S_R L_p / D_p$.

On the other hand, when the base thickness is increased, the value of Q_{st} becomes more sensitive to the intensity and direction of the built-in field than to the value of S_R.

When $W/L_p \approx 1$ (which is typical of real planar diodes) and $| E_n | = 2.5$, the change of the direction of the built-in field alters the stored charge by approximately an order of magnitude.

§ 19. FIRST (RECOVERY) PHASE

A built-in field alters condiderably the duration of first (recovery) phase during the switching of a diode from the forward to the reverse direction. The practical importance of this effect has two aspects. The reduction of the recovery phase duration t_1 by an accelerating field naturally reduces the total duration of the transient process and, therefore, increases the speed of response of a diode. However, the actual improvement is not very great for two reasons. First, the total switching time of the diode is usually the sum of two approximately equal time constants t_1 and t_2 representing the two (recovery and reverse) phases. Therefore, even if t_1 were to be reduced to zero (but t_2 remained constant), this would improve the response only by a factor of 2 or 3 such an improvement could be achieved more simply in other ways (for example by the reduction of the nonequilibrium carrier lifetime in the base). Secondly, to establish an accelerating field in the base, the concentration of donors near the p-n junction must be much higher than that in the body of the base. The calculations of Muto and Wang [108] demonstrated that in order appreciably to reduce the response time of a diode, it is necessary to satisfy the condition $N_d(0)/N_d(W) > 500$. The establishment of a high donor concentration near a p-n junction makes it difficult or altogether impossible to achieve a reasonably high value of the breakdown voltage and a small barrier capacitance.

For these two reasons, an accelerating field in the base is not widely used to reduce the semiconductor diode response time.

A retarding field (§17) slows down somewhat the transient process by increasing the value of t_1 but it reduces strongly the value of t_2. The nature of the time dependence of the transient reverse current is then described by a nearly rectangular curve.

Such devices* are being increasingly used in the amplification and formation of nanosecond pulses and in the efficient multiplication of frequency in the gigahertz range.

For this reason most of the investigations of transient processes in diodes with a built-in field have been concerned with retarding fields.

An approximate calculation of the duration of the first (recovery) phase t_1 for a diode with a retarding field in the base can be carried out by solving the equation describing the rate of change of the excess hole charge in the base:

$$\frac{dQ(t)}{dt} = i(t) - \frac{Q(t)}{\tau_p},$$ (5.13)

where i(t) is the current through the p–n junction.

This equation can be deduced by integrating Eq. (1.11) over the whole volume of the base using the boundary condition (5.5).

If the forward current before switching has been flowing for an infinitely long time, a steady–state hole distribution is established in the base and the initial condition for the integration of Eq. (5.13) is the relationship

$$Q(0) = i_f \, \tau_p.$$ (5.14)

However, if the switching takes place after a short forward current pulse of duration t_f, the initial condition, obtained by integration of Eq. (5.13), becomes

$$Q(0) = i_f \, \tau_p [1 - \exp(-t_f / \tau_p)].$$ (5.15)

We shall assume that the retarding field "pushes" holes toward the p–n junction so strongly that the excess charge remaining in the base at the end of the first (recovery) phase is negligibly small, i.e., we can assume that

$$Q(t_1) = 0.$$ (5.16)

---

* They are known as snap-off (step-recovery) charge-storage diodes.

Integration of Eq. (5.13), using the condition (5.16), gives the following expression [105] for the switching from steady-state conditions:

$$t_1 = \tau_p \ln \left(1 + \frac{i_f}{i_0} \right), \qquad (5.17)$$

and for the switching after a short forward current pulse, when the initial condition is (5.15), we obtain:

$$t_1 = \tau_p \ln \left[1 + \frac{i_f}{i_0} \left(1 - e^{-\frac{t_f}{\tau_p}} \right) \right]. \qquad (5.18)$$

If the duration of the forward current pulse is less than the hole lifetime ($t_f \ll \tau_p$), Eq. (5.18) assumes the following form for comparable values of i_f and i_0:

$$t_1 = t_f \, \frac{i_f}{i_0} = \frac{t_f}{B}. \qquad (5.19)$$

It is interesting to note that Eq. (5.17) is identical with Eq. (3.54), which has been derived for a diode with a thin base and a noninjecting ohmic contact.

We are justified in neglecting the residual charge [this has been done in the derivation of Eqs. (5.17) and (5.18)] as long as the duration of the forward current pulse is not too short and as long as it exceeds the duration of the second switching phase t_2. If these conditions are not satisfied, we must carry out a more rigorous calculation which takes into account the fact that at the moment $t = t_1$ the hole charge in the base may represent a considerable fraction of the initial stored charge.

The solutions of the equation of continuity for this case, discussed in [103], are shown graphically in Fig. 5.4. It is evident from this figure that in the presence of a retarding field the duration of the first (recovery) phase is longer and in the presence of an accelerating field is shorter than in a field-free diode. When the switching conditions are constant (B = const), the value of t_1 decreases monotonically with increasing intensity of an accelerating field, and increases monotonically with increasing intensity of a retarding field, approaching asymptotically the value corresponding to $|E_n| \gg 1$ and given by Eq. (5.17).

Analysis of the results obtained shows also that when we use short forward current pulses, the criteria of strong and weak fields include also the duration of the forward current pulse t_f. When $t_f < (4/D_p)(kT/qE)^2$, a retarding field has no effect whatever on the nature of the switching transient in a diode. The result obtained is explained qualitatively by the fact that immediately after the beginning of the forward current pulse the motion of holes takes place mainly under the action of diffusion forces and not under the action of the field; this follows from the fact that the hole density near the p–n junction is very low but the hole density gradient is high. When the forward current pulse is long ($t_f \gtrsim \tau_p$), a retarding field can be regarded as weak and it can be ignored when $|E_n| \ll 1$, but it should be regarded as strong when $|E_n| \gg 1$, i.e., the strong and weak field criteria are the same as for a steady-state distribution of holes in a diode with a retarding field.

A numerical calculation of the duration of the first (recovery) phase during the switching from steady-state conditions was car-

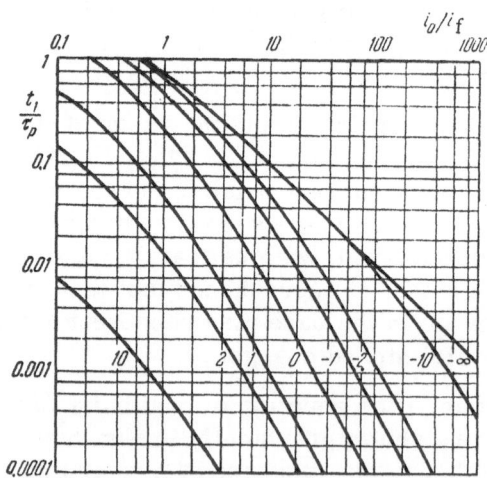

Fig. 5.4. Dependence of the duration of the first (recovery) phase on the switching conditions for diodes with different values of the built-in field E_n in the base.

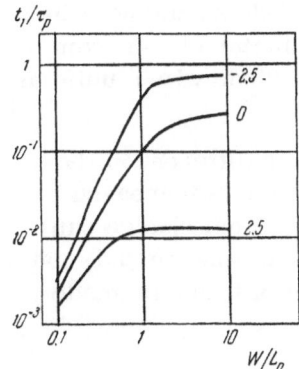

Fig. 5.5. Dependence of the duration of the first (recovery) phase on the base thickness for $i_0/i_f = 1$, $S_R L_p/D_p = 10^2$, and various values of the built-in field E_n.

ried out by Kennedy [107] for a diode with a base of arbitrary thickness. The results of his calculations for a particular value of the surface recombination velocity in the ohmic contact plane ($S_R L_p/D_p = 10^2$) are given in Fig. 5.5.

It is evident from this figure that a transition from a strong retarding field to an accelerating field may reduce the duration of the first (recovery) phase by one or two orders of magnitude. When the base thickness is reduced, the influence of the built-in field on the value of t_1 becomes weaker because the recombination of excess holes in the ohmic contact plane begins to play an increasingly important role in the hole dispersal process.

§ 20. REVERSE CURRENT DECAY

In the preceding section, we have calculated approximately the time constant t_1 for a diode with a retarding field assuming that the whole stored charge is dispersed during the first (recovery) phase. In fact, some fraction of the stored charge remains in the diode base after the end of the first phase, so that the duration of the second phase is not equal to zero.

The results of the numerical calculations made by Kennedy [107] are shown in Fig. 5.6. We can see that, when $i_0/i_f = 1$, at the end of the first phase a base with an accelerating field retains a charge of about 0.8 Q_{st}, a uniform base retains 0.4 Q_{st}, and a base with a retarding field retains less than 0.03 Q_{st}. For other values of the built-in field intensity, the residual charge (Q_{res}) in a diode with an accelerating field may reach Q_{st} (when $E \rightarrow \infty$), i.e., it may be 2–2.5 times larger than the residual charge retained in a diode with a uniform base. Assuming that an accelerating field pulls holes away from the p-n junction during the charge dispersal process, we may conclude that the duration of the second (reverse)

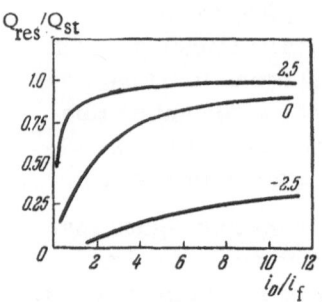

Fig. 5.6. Dependence of the residual charge Q_{res} on the switching conditions in diodes with various values for a built-in field in the base E_n (these values are given alongside the curves).

phase t_2 in such a diode should not differ more than by a factor of 2-3 from the value of t_2 for a diode with a uniform base.

The situation is different in the case $E < 0$. When $|E|$ is increased without limit, the residual charge can be made as little as we please and, consequently, the value of t_2 can be made as short as we like.

To find the time dependence of the reverse current during the second switching phase, it is necessary to solve Eq. (1.11) with the boundary condition of the (1.20) type at the p-n junction, since at a time $t = t_1$ the density of holes near the p-n junction decreases to zero. Analysis given in [103] for the case of a strong retarding field yields the following expression [t_2 is defined as the time taken by the reverse current to decay to $i(t_2) = 0.1\,i_0$]:

$$t_2 \simeq \left(\frac{kT}{q}\right)^2 \frac{2}{D_p E^2}. \qquad (5.20)$$

This formula is valid as long as the duration of the forward current pulse preceding switching is not too short [$t_f > (\tau_p/E_n^2)$]. When the forward current pulse is very short, the value of t_2 is completely independent of the diode type but is governed only by the switching conditions, i.e., by the parameters t_f, i_0, and i_f.

An approximate calculation of t_2 for the case of a retarding field has also been reported in [105]. The solution of Eq. (1.11) for the first switching phase (recovery phase) shows that at the end of this phase the hole distribution in the base is of the form

$$p(x) = \frac{j_0}{q D_p}\, x \exp\left(-x\,\frac{qE}{kT}\right). \qquad (5.21)$$

Since the duration of the second switching phase (reverse phase) t_2 is considerably shorter than the hole lifetime in diodes

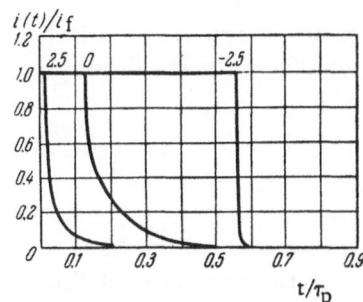

Fig. 5.7. Oscillograms of the transient reverse current after switching for three diodes with $i_0/i_f = 1$, $W/L_p = 1$, $S_R L_p/D_p = 10^2$, and various values of E_n (these values are given alongside the curves).

with a strong retarding field, it is quite permissible to ignore the recombination losses in the determination of t_2. Then, we may assume that the total hole charge remaining in the base at the beginning of the second (reverse) phase is extracted by the transient current, i.e., the following relationship is satisfied

$$\int_0^\infty i(t)\,dt = qS \int_0^\infty p(x)\,dx. \quad (5.22)$$

We shall assume that the reverse current decay during the second (reverse) phase is described by the exponential law:

$$i(t) \simeq i_0 e^{-t/t_{dec}} \quad (5.23)$$

Substituting Eqs. (5.21) and (5.23) into Eq. (5.22), we obtain

$$t_{dec} = \frac{1}{D_p}\left(\frac{kT}{qE}\right)^2 \quad (5.24)$$

and, consequently,

$$t_2 = \left(\frac{kT}{q}\right)^2 \frac{2.3}{D_p E^2}, \quad (5.25)$$

which is practically identical with Eq. (5.20).

The amounts of change recovered during the first and second phases of a switching transient are given by $Q_{rec1} = i_0 t_1$ and $Q_{rec2} = i_0 t_{dec}$, respectively [the value of Q_{rec2} is found by integrating Eq. (5.23)]. The ratio of these charges, representing the rate of recovery of the reverse resistance in a diode with a strong retarding field, is obtained from Eqs. (5.17) and (5.20):

$$\frac{Q_{rec2}}{Q_{rec1}} = \left[4E_n^2 \ln\left(\frac{1+B}{B}\right)\right]^{-1}, \quad (5.26)$$

which simplifies to the following expression for large values of B (which are typical of some applications of charge-storage diodes):

$$\frac{Q_{rec2}}{Q_{rec1}} = \frac{i_0}{4i_f \, E_n^2}.$$

(5.27)

Thus, the higher the normalized field intensity in the base and the lower the ratio $B = i_0/i_f$, the closer is the approach of the transient response of the reverse current to the rectangular shape.

The influence of a built-in field on the transient character-istic of the reverse current after switching can be seen from os-cillograms calculated by Kennedy [107] and shown in Fig. 5.7. The effect of a built-in field must be allowed for when the hole lifetime in diodes with a nonuniform base is measured using one of the me-thods based on the switching transient.

Chapter VI

Transient Processes in Diodes During the

Passage of a Forward Current Pulse

§ 21. INTRODUCTION

We have established that when a forward bias is applied to a diode, an equilibrium distribution of free carriers in the base is not established instantaneously but after a certain time interval.

Thus, the transient process of the establishment of an equilibrium distribution is responsible for the time dependences of the measurable electrical characteristics of a diode (the voltage at its external terminals or the current through it). Since the forward resistance of a diode is low, it is experimentally more convenient to investigate the process of the establishment of an equilibrium forward state under current generator conditions. The simplest circuit for the observation of this transient process is shown in Fig. 6.1a. The application of rectangular pulses of large resistance (exceeding considerably the forward resistance of the diode) between the generator and the diode produces rectangular current pulses suitable for investigating the diode behavior.

It was demonstrated in § 3 that, at low injection levels, the distribution of the excess hole density in the base at various moments after the beginning of the flow of the forward current is given by Eq. (1.55) and the graphical dependences described by this equation are given in Fig. 1.3 for several values of the parameter $\mathcal{J}_f$. From the curves in Fig. 1.3, as well as from Eq. (1.58), which gives the impressed hole density near the junction, it is clear that at the moment of application of a current pulse the density of holes

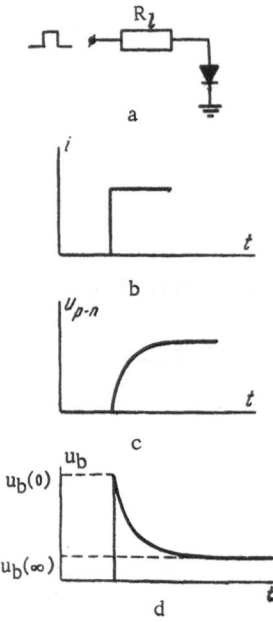

Fig. 6.1. Transient process of the establishment of a forward voltage across the diode: a) diode circuit; b) current pulse; c) time dependence od the voltage drop across the p-n junction; d) time dependence of the voltage drop across the base.

near the p-n junction is equal to its initial value p_{n0} and it reaches its new equilibrium value p_1 sometime after the beginning of the current pulse.

Physically, it is self-evident that an exactly identical monotonic rise of the impressed hole density at the p-n junction from the value p_{n0} to p_1 will be observed in the case of an arbitrary injection level for a diode with an arbitrary geometry of its p-n junction and its base. All that changes is the quantitative relationship, governing the rate of rise, p(t), but the value of p_1 has its own specific value in every particular case.

According to Eq. (2.47), the voltage across the p-n junction increases during this process from zero to a maximum value corresponding to the flow of a steady-state forward current, equal to the amplitude of the current pulse. The time dependence of the voltage across the p-n junction after the application of a forward current step is shown in Fig. 6.1c. We can see that when the p-n junction is connected in the forward direction, it behaves like a capacitor. The capacitative reaction of the p-n junction to external signals is observed also when a forward current pulse is applied to it. In fact, according to Eq. (2.74), the forward current under such conditions decreases from an infinitely high value at the moment of switching to some steady-state level, i.e., it behaves exactly as a transient current through a capacitor when a voltage step is applied to it. We note that both Eqs. (2.47) and (2.74) have been obtained neglecting the electrostatic capacitance of the p-n junction and they describe only changes in the excess hole distribution in the base.

So far, considering the various transient processes in diodes, we have usually neglected the voltage drop across the distributed base resistance because in the case of a reverse bias this voltage drop is small compared with the drop across the p–n junction. The situation is different in the case of the transient process of the establishment of an equilibrium forward state. In this case, the voltage drop across the base may be comparable with the drop across the p–n junction (which, in any case, is not greater than the value of φ_0) and at high currents it may exceed considerably the value of u_{p-n}.

After the application of a forward current step to a diode the voltage drop across the base u_b decreases (due to the modulation of the resistance of the semiconductor by injected carriers) from some initial value to a value corresponding to an infinitely long forward current pulse. Initially, the base resistance is governed only by the geometrical dimensions and the resistivity of the semiconductor material. During the flow of the forward current, the density of the mobile carriers in the region next to the p–n junction increases: the junction injects holes and, to ensure the condition of neutrality, the positive charge of holes is compensated by the arrival of the same number of electrons from the ohmic contact. Thus, the resistivity of the part of the base near the p–n junction during the flow of the forward current is less than that in the absence of such a current and therefore the total resistance of the base decreases. The time dependence of the voltage drop across the base resistance is shown graphically in Fig. 6.1d. We can see from that figure that the base, in contrast to the p–n junction, has an inductive reaction to an external electric signal. The difference between the values $u_b(0)$ and $u_b(\infty)$ increases with increasing density of the forward current through the p–n junction. In general, the effect of modulation of the base resistance becomes appreciable only under conditions such that the density of injected holes becomes comparable with or greater than the value n_{n0}, i.e., when the injection level is moderate or high.

From this discussion it is clear that the reaction of a diode (which represents a series-connected p–n junction and the base resistance) to a forward current step depends on the amplitude of this step. At low values of the current, when the voltage drop across the base is negligibly small compared with the drop across the p–n junction, the diode behaves as a capacitor: the voltage across it gradually increases from some initial value to a steady-state level.

When the density of the forward current is very high, so that the processes in the base become dominant, a diode behaves like an inductor: an initial forward voltage peak is followed by a gradual decrease of the voltage.

We must stress that the appearance of inductive properties in a diode is possible only because the external voltage is divided between the p-n junction and the base. If all the applied voltage is concentrated in the p-n junction, the diode behaves like a capacitor; if it is concentrated only in the base, the diode behaves like an ohmic resistor.

There have been may investigations of the inductive properties of forward-biased diodes, which appear during the application of a small sinusoidal signal (cf., for example, [109-115]), but we shall not consider these properties in detail. We shall only discuss the transient processes of the establishment of a forward voltage which occompanies a step-like rise of the forward current through a diode from zero to some given value.

The solution of this problem is meaningful only for a planar diode with a base which is limited in area and in thickness, or for a diode with a hemispherical p-n junction and an infinite base. Only for these two diode models is the base resistance finite.

The inductive properties of diodes have been pointed out in the earliest investigations of transient processes in semiconductor rectifiers, simultaneously with the discovery of their slow response during the switching from the forward to the reverse direction [116, 10, 117]. Subsequent numerous investigation [118-120, 52, 121-123, 71] have demonstrated that this effect is due to the injection of holes by the p-n junction and the resultant modulation of the base resistance. Several important qualitative and semiquantitative relationships have been established but even now formulas as accurate and clear as those for the switching from the forward to the reverse direction are not available for the transient process following the application of step-like forward current to a neutral diode. The reason for this is that the inductive properties of diodes are appreciable only at high injection levels when the equation of continuity cannot be expressed in an exact analytic form.

§ 22. ESTABLISHMENT OF A FORWARD
RESISTANCE IN A PLANAR DIODE

We shall discuss the transient processes accompanying the passage of a forward current pulse through a diode, making the following assumptions: the p-n junction is abrupt and asymmetrical (the conductivity of the p-type region is considerably higher than the conductivity of the n-type region); at any point in the n-type region, the condition of electrical neutrality is satisfied at every moment; the influence of trapping levels on the transient processes can be neglected (cf. Chap. VII); investigated diodes can be described by a model of a planar diode with a thin base and an ideal recombination-type ohmic contact ($S_R = \infty$). In contrast to the general planar diode model, we shall assume the p-n junction area to be finite (since only in this case the base resistance is not equal to zero), but we shall postulate that the distortions in the lines of the current at the edges of the base are slight, i.e., that the problem is still one-dimensional.

When these assumptions are made, the behavior of holes in the base of a diode is described by Eq. (1.15) and the field intensity at each point in the base is given by Eq. (1.16). Since we have assumed an infinite surface recombination velocity in the ohmic contact plane, the boundary condition at x = W is described by Eq. (3.2). The boundary condition at the p-n junction, after the transformation given by Eq. (1.17), is obtained in the form

$$\left.\frac{\partial p}{\partial x}\right|_{x=0} = -\frac{J_f}{qD_p}\left[\frac{2\delta p + n_{n0} + p_{n0}}{\delta p + n_{n0}}\right]_{x=0}. \tag{6.1}$$

Equation (1.15) with the boundary conditions given by the nonlinear equation (6.1) and by Eq. (3.2) can be solved only by numerical integration.

Kano and Reich [124] used a digital computer to find the solution for an alloyed silicon diode with an n-type base of $\rho_n = 20~\Omega\cdot cm$ resistivity and $\tau_p = 7~\mu sec$. The results of their calculations are presented graphically in Figs. 6.2-6.4. It is evident from Fig. 6.2 that the voltage drop across the p-n junction is established in a time interval which is much shorter (by at least one order of

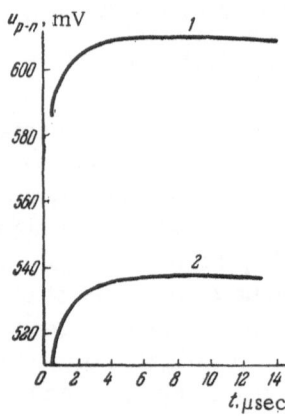

Fig. 6.2. Establishment of a forward voltage across a p-n junction for current densities (A/cm^2): 1)10; 2) 0.5.

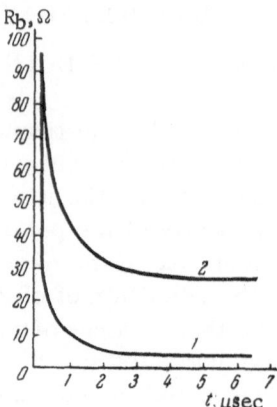

Fig. 6.3. Establishment of a base resistance for forward pulse current densities (A/cm^2); 1) 10; 2) 0.5.

magnitude) than the hole lifetime in the base. It is arbitrarily assumed that the transient process is complete when the value of u_{p-n} reaches 0.9 of its steady-state level. We can see also that the higher the forward current density, the more rapid is the establishment of the value u_{p-n}.

The dependence of the base resistance on time (Fig. 6.3) shows that the duration of the resistance recovery process is also shorter than τ_p but it is considerably longer than the time taken to establish the voltage across the p-n junction. When the current density is increased, the base resistance decreases more rapidly and to a lower steady-state level.

The time dependence of the total voltage across the external terminals of a diode may obey three different laws, depending on the forward current density. At low forward current densities, the voltage drop across the p-n junction exceeds the voltage drop across the base and therefore the transient process of the establishment of the voltage across the diode is practically identical with the establishment of the voltage across the p-n junction: the diode behaves like a capacitor. At high forward current densities, the situation is reversed, $u_b > u_{p-n}$; therefore, the process of the

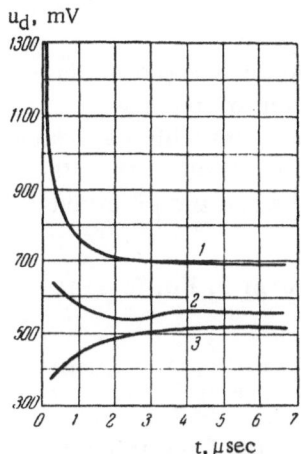

u_d, mV

t, μsec

Fig. 6.4. Transient process of the establishment of a forward voltage across a diode for current densities (A/cm^2): 1) 10; 2) 0.5; 3) 0.05;

establishment of the voltage across the diode is governed primarily by the change in the base resistance: the diode behaves like an inductor.

In the intermediate case of moderate current densities, the process of the establishment of the forward voltage across the diode is oscillatory (Fig. 6.4). The oscillations are due to the fact that at moderate current densities the rise of the voltage across the p-n junction is approximately compensated by the fall of the voltage drop across the base; however, total compensation is not obtained at all times because the time dependences $u_{p-n}(t)$ and $u_b(t)$ are different.

Some writers refer to the critical current density at which the diode behaves like an ohmic resistance. It is obvious that, strictly speaking, such a situation is impossible to reach (because of oscillations) and the critical condition should be regarded as the state when neither the capacitative nor the inductive reactions of the diode predominate throughout the transient process.

Experiments carried out on many diodes have confirmed the correctness of the numerical results of Kano and Reich. Additional estimates have shown that the nature of the transient process of the establishment of the voltage after a forward current pulse is governed by such parameters of the diode as the resistivity of the base material and the base thickness. The value of the hole life-time has much less effect.

As mentioned in § 2, Eq. (1.15) can be simplified to Eq. (1.13) when the injection is level is very low, or to Eq. (1.18) when the injection level is very high. In such cases, the electric field intensity in the base is described by Eq. (1.16). The total voltage drop across the diode is

$$u_d = u_{p-n} + \int_0^w E\,dx = u_{p-n} + \int_0^w E_1\,dx + \int_0^w E_2\,dx = u_{p-n} + u_0 + u_D, \quad (6.2)$$

where E_1 and E_2 are the first and second terms on the right-hand side of Eq. (1.16).

The quantity u_O is the ohmic voltage drop across the modulated base resistance during the flow of a current of density j through the base; the quantity u_D is the Dember emf which appears in the interior of a semiconductor in all those cases when the distribution of excess carriers is nonuniform and when the electron and hole mobilities are unequal.

It follows from Eq. (1.21) that the first term in Eq. (6.2) is

$$u_{p-n}(t) = \frac{kT}{q} \ln\left[1 + \frac{\delta p\,(0,\,t)}{p_{n0}}\right]. \tag{6.3}$$

Integration of the Dember field intensity across the base thickness, taking into account the fact that $\delta p(W) = 0$, yields

$$u_D(t) = \frac{kT}{q}\frac{b-1}{b+1} \ln\left[1 + \frac{(b+1)\,\delta p\,(0,\,t)}{b n_{n0} + p_{n0}}\right]. \tag{6.4}$$

Finally, the ohmic voltage drop can be found by integrating the expression $\int_0^W E_1\,dx$ only if the specific equation for $\delta p(x, t)$ is substituted in it. Thus, to determine the transient process of the establishment of a forward voltage across a diode it is necessary to know the function $\delta p)x, t)$.

The solution of Eq. (1.13) with the boundary conditions (3.2) and (6.1) has been obtained by Carslaw and Jaeger [20] for the conduction of heat in solids. The functions $\delta p(x, t)$, valid at any moment after the application of a forward current pulse, are described by the fairly cumbersome formulas given in [125]. When these formulas are used, the value of $u_O(t)$ can be found only by numerical integration methods.

The agreement between the calculated and experimental data can be seen in Fig. 6.5, which gives curves obtained by Chang [125]. Chang investigated alloyed germanium diodes with the following parameters: $p_{p0} = 2 \cdot 10^{18}$ cm^{-3}, $n_{n0} = 7 \cdot 10^{13}$ cm^{-3}; $W = 300\,\mu$; $S = 10^{-2}$ cm^2. In these diodes, the volume lifetime in the base was so high that recombination during the transient process of the

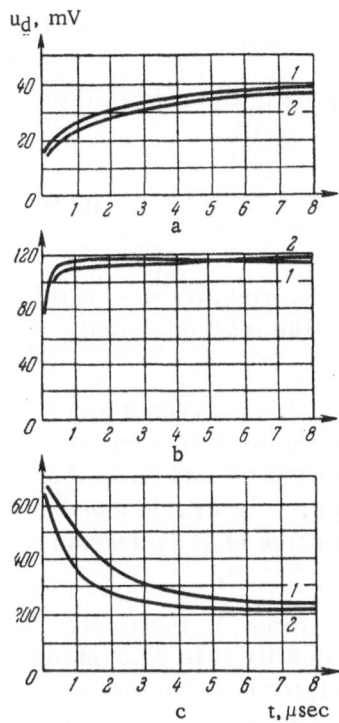

u_d, mV

Fig. 6.5. Transient process of the establishment of a forward voltage across a diode for current amplitudes (mA): a) 0.1; b) 1; c) 10. 1) Calculated curve; 2) experimental curve.

establishment of a forward voltage could be neglected. Examination of the curves in Fig. 6.5 shows good agreement between the calculated and experimental results, particularly for the forward currents of 0.1 and 1 mA. The agreement for 10 mA can also be regarded as satisfactory.

The curves in Fig. (6.5) show also that when i_f = 0.1 mA, a diode exhibits a capacitative reaction, but when i_f = 10 mA it exhibits an inductive reaction. The value i_f = 1 mA is close to the critical value because then the diode behaves almost like an ohmic resistor.

The general method for finding the critical value of the current i_{cr} is as follows. At the moment of switching, the voltage drop across the p-n junction and the Dember emf in the base are equal to zero, as indicated by Eqs. (6.3) and (6.4). The voltage drop across the diode at this moment is exactly equal to the voltage drop across an unmodulated base resistance

$$R_b(0) = \frac{W\rho_n}{S} .   (6.5)$$

The value $R_b(0)$ can be found experimentally by measuring the maximum forward voltage drop across the diode for a known current pulse amplitude.

If the current pulse amplitude is equal to the critical value, the voltage drop across an unmodulated base resistance at the moment of switching should be equal to the steady-state voltage drop across the diode. Thus, the value of i_{cr} can be found graphically from the point of intersection of the straight line $R_b(0)$ with the static current-voltage characteristic of the diode (Fig. 6.6). An analytic expression for i_{cr} is:

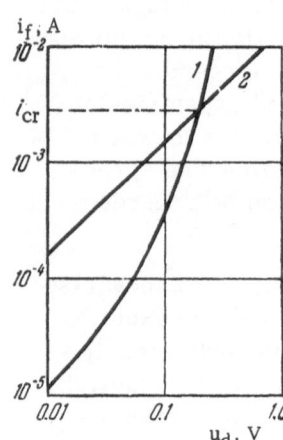

Fig. 6.6. Graphical method
for determination of the critical current. 1) Static current-voltage characteristic
of a diode; 2) $u_d(0) = f(i_f)$
line.

$$i_{cr} = 2 \frac{kT}{q} \frac{\varkappa}{R_6(0)}, \qquad (6.6)$$

where the coefficient $\varkappa$ is the root of the equation

$$e^{\varkappa} = \varkappa \left(\frac{b+1}{b} \right)^{1/2} \left[\frac{(b+1) \rho_i \tanh(W/L_p)}{\rho_n (W/L_p)} \right]. \qquad (6.7)$$

Here, ρ_n and ρ_i are the resistivities of the n-type region and of the intrinsic semiconductor.

The value of $\varkappa$ lies between 3 and 6 in those cases when the base is made of germanium, and between 11 and 15 in the case of silicon bases. When gold-doped silicon is used, the value of $\varkappa$ decreases to 8–12.

Thus, assuming that, in the first approximation, the quatity $\varkappa$ is a constant for a given material, we find that the critical current is inversely proportional to the unmodulated base resistance $R_b(0)$.

In spite of the fact that numerical calculations based on the exact equation (1.15) or the approximate form (1.13) make it possible to find the transient response with a high degree of accuracy, the results are not very interesting because general relationships for different types of diode cannot be deduced by numerical computation.

For a diode with a sufficiently thin base ($W \lessgtr L_p$), in which a high injection level applies throughout the base, so that we can use Eq. (1.18), we can obtain a general expression which describes the behavior of the forward voltage drop across the diode during the transient process considered here.

The boundary condition in the p-n junction plane, given by Eq. (6.1), assumes the following form when $\delta p \gg n_{n0}$:

$$\frac{\partial p}{\partial x} \bigg|_{x=0} = -\frac{2j_f}{qD_p}, \qquad (6.8)$$

which differs only by a factor 2 from the condition (1.20) which is valid at low injection levels. Physically, this means that at very high injection levels the hole drift current through the p-n junction is equal to the diffusion current.

The solution of Eq. (1.18), with a zero initial distribution of the excess hole density and with the boundary conditions given by Eqs. (3.2) and (6.8), is given by Eq. (3.46), as demonstrated earlier.

Since the infinite sum in this expression for $W_n \lesssim 1$ is a rapidly converging series, we need consider only the zeroth term after a certain time from the beginning of the forward current pulse.

The time dependence of the injected hole density near the p-n junction (x = 0) is described by the following expression

$$\delta p(0, t) \simeq p_{n0} \frac{j_f}{2j_s}[1 - m \exp(-t/\tau_{W0})],\qquad(6.9)$$

where

$$j_s = \frac{qD_p p_{n0}}{L_p} \coth\frac{W}{L_p},\qquad(6.10)$$

$$\frac{1}{\tau_{W0}} = \frac{1}{\tau_p}\left(1 + \frac{\pi^2}{4W_n^2}\right),\qquad(6.11)$$

$$m = \frac{2W_n \coth W_n}{\pi^2/4 + W_n^2}.\qquad(6.12)$$

Here, j_s is, as indicated by Eq. (3.9), the reverse saturation current in a thin-base diode. In the definition of the quatities τ_{W0}, j_s, and m we have assumed that L_p is the value of this quantity at high injection levels [cf. discussion following Eq. (1.18) in § 2].

Combining Eqs. (1.21) and (6.9), we find that the time dependence of the voltage across the p-n junction during the switching transient is given by the following expression:

$$u_{p-n}(t) = \frac{kT}{q} \ln\left\{\frac{j_f}{2j_s}[1 - m \exp(-t/\tau_{W0})]\right\},\qquad(6.13)$$

which is valid provided the values of t/τ_{W0} are not too low.

Using Eq. (1.16), the voltage drop across the diode base is found to be

$$u_b = \int_0^W E\,dx = \frac{j_f}{q\mu_p\,(b+1)} \int_0^W \frac{dx}{(b+1)\,p\,(x) + bn_{no}} -$$

$$- \frac{kT}{q}\frac{b-1}{b+1}\ln\frac{bn_{no}}{(b+1)\,p\,(0) + n_{no}}. \qquad (6.14)$$

As before, the first term represents the ohmic voltage drop across the base and the second term is the Dember emf. The structure of Eq. (6.14) shows that the Dember emf u_D vanishes when $\mu_n = \mu_p$ (b = 1).

When the injection level is high throughout the base and $(W/L_p) \lesssim 1$, the integral in Eq. (6.14) can be determined approximately. Afer making several simplifying transformations, the general expression for the time dependence of the forward voltage across the diode during the passage of a current pulse of density j_f assumes the following form [55]:

$$\frac{qu\,(t)}{kT} = \ln\left\{\frac{j_f}{2j_s}\,[1 - m\exp(-t/\tau_{w0})]\right\} +$$

$$+ \frac{2\cosh W_n}{(b+1)\,[1 - \exp(-t/\tau_{w0})]}\ln\left\{\frac{j_f}{2j_{s1}}\,[1 - \exp(-t/\tau_{w0})]\right\} +$$

$$+ \frac{b-1}{b+1}\ln\left\{\frac{j_f}{2j_{s2}}\,[1 - m\exp(-t/\tau_{w0})]\right\}, \qquad (6.15)$$

which is valid provided the values of t/τ_{w0} are not too low. Here, j_s, τ_{w0}, and m are given by the relationships (6.10)–(6.12), and

$$j_{s1} = j_s\,\frac{n_{no}}{2p_{no}\,(b+1)}\,\frac{\sinh W_n}{\tanh W_n/2}, \qquad (6.16)$$

$$j_{s2} = j_s\,\frac{n_{no}}{p_{no}\,(b+1)}. \qquad (6.17)$$

We can easily see that at low values of W_n, the current densities j_{s1} and j_{s2} are equal. Let us analyze Eq. (6.15). First of all, we note that when $t \rightarrow \infty$, we obtain the relationship between the current and the voltage across the diode which was established earlier by Stafeev [16] for the steady-state case.

After the application of a forward current step to a diode, the voltage drop across the p-n junction and the Dember emf in the base increase from zero to their steady-state values. At the same time, the ohmic voltage drop across the base decreases monotonically. This ohmic voltage drop varies more rapidly with time than the other two terms, since they have time-dependent coefficients only in the logarithms.

All three components of the voltage drop across the diode have the same time constant τ_{W_0}. Since for real diodes, we always have $j_{S2} \gg j_S$, it follows that the Dember potential difference is much smaller than the voltage drop across the space-charge layer.

In the first approximation, we may assume that the total voltage drop across the diode consists of the drop across the p-n junction and the ohmic drop across the base. If the forward current is sufficiently large, the steady-state value of u_{p-n} is established almost instantaneously and the dependence $u_d(t)$ is governed primarily by the time dependence of the ohmic voltage drop across the base, i.e., by the modulation of the base resistance.

§ 23. ESTABLISHMENT OF A FORWARD VOLTAGE ACROSS A DIODE WITH A HEMISPHERICAL p-n JUNCTION

Armstrong and his colleagues [102, 126] have considered the relationship between the forward current and the applied voltage under transient and steady-state conditions for diodes with hemispherical p-n junction.

We shall make the usual assumptions (§ 14) about the model of a diode with a hemispherical p-n junction: the lines of the current are radial; the p- and n-type regions have different doping levels; the electrical neutrality applies at any point in the base at every moment.

In contrast to § 14, we shall consider the case of high injection levels because this case represents the conditions in the majority of experiments on diodes having small-area p-n junctions.

The relationship between the hole and electron currents, on the one hand, and the free carrier density and electric field intensity, on the other, is described by equations similar to the system (1.3)-(1.8) provided the functions in that system are assumed to depend on three coordinates and the derivative $\partial/\partial x$ is replaced by the operator ∇. Then, transformations similar to those described in §2 yield the diffusion equation for holes in the form of Eq. (4.1) in which D_p is replaced with D' in accordance with the relationship

$$\frac{1}{D'} = \frac{1}{D_n} + \frac{1}{D_p}. \qquad (6.18)$$

At very high current densities, when the injected carrier density near the p-n junction is comparable with p_{p_0}, we cannot assume that the injection efficiency is equal to unity. If we assume that the carrier lifetime is the same on both sides of the p-n junction, the expression for the injection efficiency becomes:

$$\gamma = \frac{\mu_p p_{p0}}{\mu_p p_{p0} + \mu_n (p + n_{n0})}. \qquad (6.19)$$

When the forward bias is high so that the potential barrier in the region of the p-n junction practically disappears, we have $p \approx p_{p0}$ and it follows from Eq. (6.19) that the current flowing through the junction from each side is proportional to the density of the majority carriers on that side (we note that $p + n_{n0} = n$) and to the majority-carrier mobility. In the case of small currents (and correspondingly small values of p), the expression (6.19) reduces to the well-known formula for γ, where $\gamma \to 1$ when $p_0 \gg n_{n0}$. Thus, we may assume that Eq. (6.19) is valid throughout the whole range of the density of injected holes from zero to p_{p0}.

Combining the three-dimensional forms of Eqs. (1.6)-(1.7) with Eq. (6.19), we obtain

$$E = \frac{p_{p0} + p + n_{n0}}{q\,(2p + n_{n0})\,[\mu_p p_{p0} + \mu_n\,(p + n_{n0})]}\, j_f\,, \qquad (6.20)$$

$$j_f = \frac{q\,(2p + n_{n0})\,[D_p p_{p0} + D_n\,(p + n_{n0})]}{(p + n_{n0})\,(p_{p0} - p)}\,\nabla p. \tag{6.21}$$

Thus, an approximate expression for γ allows us to obtain the relationship between E and j_f in the form of direct proportinonality, which is simpler than the general formula (1.16). It follows from Eq. (6.21) that the potential barrier in the region of the p–n junction cannot be suppressed completely since, in this case, we would have $p \to p_{p0}$ and $j_f \to \infty$.

At very high forward current densities, we can substitute approximately $p(r_0) \approx p_{p0}$ into the numerator of Eq. (6.21). Then, using the simple relationship $i_f = j_f \cdot 2\pi r_0^2$ for the current through the rectifying contact, we obtain

$$i_f \simeq -\frac{8\pi r_0^2 q D'' p_{p0}}{p_{p0} - p\,(r_0)}\,\nabla p\Big|_{r=r_0}. \tag{6.22}$$

Here, D" represents the arithmetic mean of the values of D_n and D_p. For simplicity, we shall assume that D" = D' = D.

The solution of Eq. (4.1) with the boundary condition (6.22), the zero initial condition, and the condition that $p_{n0} \approx 0$ at infinity, can be obtained by the operator method. The solution gives the following expression for the time dependence of the hole density near the p–n junction:

$$p\,(r_0) = p_{p0}\Big(A - \frac{\sqrt{D}}{r_0}\Big)\Big(\frac{1}{A^2 - 1/\tau_p}\Big)\Big\{ A - \frac{1}{\sqrt{\tau_p}}\,\mathrm{erf}\,\sqrt{\frac{t}{\tau_p}} -$$

$$- A\exp{[(A^2 - 1/\tau_p)\,t]}\,\mathrm{erfc}\,(A\sqrt{t})\Big\}, \tag{6.23}$$

where

$$A = \frac{i_f}{8\pi r_0^2 q p_{p0}\sqrt{D}} + \frac{\sqrt{D}}{r_0}. \tag{6.24}$$

The voltage drop across the base resistance during the flow of the forward current is

$$u_b = \int\limits_{r_0}^{\infty} E\,(r)\,dr. \tag{6.25}$$

We shall now determine the function $E(r)$. It follows from our assumption of the radial nature of the lines of the current that the lines of force of the electric field are also radial, i.e., E is a function only of the radius.

When the charge neutrality applies at every point in the base, Eq. (1.8) can be expressed in terms of spherical coordinates:

$$\operatorname{div} E = 0 = \frac{1}{r^2} \frac{\partial}{\partial r} [r^2 E(r)]. \tag{6.26}$$

The solution of this equation is

$$E(r) = \frac{r_0^2}{r^2} E(r_0). \tag{6.27}$$

Substituting Eq. (6.27) into Eq. (6.25), we obtain

$$u_b \simeq r_0 E(r_0). \tag{6.28}$$

To simplify out calculations, we shall assume that the electron and hole mobilities are equal and independent of the respective carrier densities, i.e., $\mu_n = \mu_p = \mu$. The assumption of the equality of the carrier mobilities is justified, as indicated by the curves in Fig. 2.17, at carrier densities of the order of 10^{18} cm^{-3}. By ignoring the difference between μ_n and μ_p, we automatically exclude the Dember emf whose value, as demonstrated in the preceding section, is only a small fraction of the total forward voltage drop across the diode.

Then, Eqs. (6.28) and (6.20) yield

$$u_b \simeq \frac{i_f}{2\pi q \mu r_0 [2p(r_0) + n_{n0}]}, \tag{6.29}$$

i.e., if the assumptions that we have made are valid, the base resistance is equal to the resistance of the hemispherical contact in a material whose electrical conductivity is

$$q\mu [2p(r_0) + n_{n0}] = q\mu [p(r_0) + n(r_0)].$$

The voltage drop across the p-n junction is given by the following expression, which applies in the case of Boltzmann's statistics:

$$u_{p-n} = \frac{kT}{q} \ln \frac{p(r_0)}{p_{n0}}. \tag{6.30}$$

If we assume that Eqs. (6.29) and (6.30) are valid also under transient conditions, the required transient characteristic u(t) can be found by substituting into these equations the solution (6.23) for $p(r_0, t)$.

We shall now make some simplifications. For a small radius of the rectifying contact ($r_0 \ll \sqrt{D\tau}$), it follows from Eq. (6.24) that $A^2 \gg 1/\tau$, and when we neglect that term $1/\tau_p$ compared with A^2, the distribution (6.23) becomes

$$p(r_0, t) \simeq \frac{i_f \, p_{p0}}{i_f + 8\pi r_0 q p_{p0} D} \left[1 - e^{A^2 t} \mathrm{erfc}\left(A \sqrt{t} \right) \right]. \tag{6.31}$$

When the amplitude of the forward current is increased, we ultimately obtain

$$\frac{i_f}{i_f + 8\pi r_0 q p_{p0} D} \to 1. \tag{6.32}$$

By way of illustration, we shall show that the relationship (6.32) is satisfied for a typical point-contact diode ($r_0 \approx 5 \cdot 10^{-4}$ cm, $p_{p0} \approx 5 \cdot 10^{16}$ cm^{-3}, $D \approx 30$ cm^2/sec) by forward currents as low as $i_f > 5$ mA.

Substituing Eq. (6.31) into Eqs. (6.29) and (6.30), and using Eq. (6.32), we obtain

$$u = \frac{i_f}{2\pi r_0 \mu q \left[2 p_{p0} f(t) + n_{n0} \right]} + \frac{kT}{q} \ln \left[1 + \left(\frac{p_{p0}}{p_{n0}} - 1 \right) f(t) \right], \tag{6.33}$$

where

$$f(t) = 1 - e^{A^2 t} \, \mathrm{erfc}\left(A \sqrt{t} \right).$$

When $p_{p0} \gg n_{n0}$, the quantity $R_r = 1/4\pi \, r_0 q \mu p_{p0}$ represents the "residual" base resistance, i.e., the lowest limit to which R_b tends when the forward current rises without limit and $p(r_0)$ approaches

p_{p0}. Noting that $(kT/q) \ln (p_{p0}/p_{n0})$ is the contact potential or equilibrium barrier height φ_0, we can transform Eq. (6.33) into

$$u(t) = \frac{R_R i_f}{\frac{n_{n0}}{2p_{p0}} + f(t)} + \varphi_0 + \frac{kT}{q} \ln \left[\frac{p_{n0}}{p_{p0}} + f(t) \right]. \qquad (6.34)$$

The function $f(t)$ increases monotonically from 0 to 1 when t is varied from 0 to ∞. Thus, after an appreciable time from the beginning of the flow of the forward current pulse, the expression (6.34) can be simplified to:

$$u(t) = \frac{R_R i_f}{f(t)} + \varphi_0 + \frac{kT}{q} \ln f(t). \qquad (6.35)$$

Under steady-state conditions $(t \rightarrow \infty)$, we obtain the following self-evident relationship:

$$u(\infty) = R_R i_f + \varphi_0. \qquad (6.36)$$

When t = 0, it follows from Eq. (6.34) that

$$u(0) = \frac{i_f}{2\pi r_0 q \mu n_{n0}}, \qquad (6.37)$$

i.e., the voltage drop across the diode is initially equal to the voltage drop across the spreading resistance of a hemisphere of radius r_0 in a semiconductor with a majority-carrier density of n_{n0}. Physically, this conclusion is reasonable because initially the voltage drop across the p-n junction is equal to zero and the base resistance is not modulated.

Chapter VII

Transient Processes in Semiconductor Diodes and Fundamentals of Recombination Theory

§ 24. INTRODUCTION

In the theory of the relaxation processes taking place in semiconductors under nonequilibrium conditions, it is usual to employ a basic characteristic in the form of the minority-carrier lifetime or, in the case of an n-type semiconductor, the hole lifetime τ_p.

We have used this parameter in discussing switching transients in diodes and, in the majority of cases, we have assumed that τ_p is constant for a given semiconducting material. In other words, we have assumed that the process of the decrease of the density of injected excess holes due to recombination is described by a simple exponential law

$$p(t) = p(0) \exp(-t/\tau_p), \qquad (7.1)$$

where τ_p is independent of the hole density and of processes (other than recombination) which tend to maintain or disturb the nonequilibrium state. The validity of this assumption is not self-evident; on the contrary, it is evident that the relaxation of the excess hole density should obey a law more complex than that given by Eq. (7.1).

24.1. Recombination Centers

From general theoretical considerations it follows that, in semiconductors such as germanium or silicon, recombination of electrons and holes in an ideal lattice is practically impossible, and that it can take place only at some structure defects which act

197

as recombination centers or traps. This is because the laws of conservation of energy and momentum must be satisfied by every recombination event. In the case of recombination at a structure defect, whose mass is infinitely large compared with the mass of an elementary particle, this defect can absorb any momentum of the captured carrier and this increases the probability of recombination at such a defect compared with the direct recombination of free electrons and holes.

Numerous experiments have confirmed the trapping mechanism of recombination of excess carriers in germanium and silicon.

A great variety of structure imperfections may act as traps. They include dislocations formed during growth of single crystals; imperfections and cracks resulting from various mechanical deformations; relative displacements of parts of the crystal lattice and defects produced by heat treatment or by nuclear radiations; departures from the periodicity of a crystal at the surface; atoms of various impurities located at sites or interstices in the host lattice (such impurities are, in fact, the most important type of trap).

In all these cases, the presence of recombination centers is manifested by the appearance of allowed energy levels in the forbidden band of a semiconductor.

The efficiency of recombination at an impurity center is governed primarily by the probability of carrier capture by the center, which depends on the electron and hole capture cross sections. The capture cross sections represent the interaction of a free carrier with a trap. A recombination center can be represented schematically in the form of a circle of area σ_p (for holes) on σ_n (for electrons): when a carrier passes through such a circle it is captured by the center. The larger the capture cross section σ_p and σ_n, the more efficient is the recombination at a given center.

24.2. Recombination Statistics for a
Singly Charged Center

Shockley and Read [13] considered the statistics of recombination capture in the simplest case when only one type of recombination center, represented by a single local energy level, is present in a crystal. Their analysis applies to steady-state processes during the uniform generation and recombination of excess carriers throughout a semiconductor. Moreover, they assumed that the concentration of traps is low compared with the density of nonequilibrium carriers; this makes it possible to neglect the charge of trapped electrons and holes and to write the neutrality condition in the form $\delta p = \delta n$. A low trap concentration means also that when the behavior of free carriers is described by the Boltzmann distribution, those carriers which are captured by traps obey the Fermi statistics. When these assumptions are made, the expression for the lifetime of nonequilibrium electron–hole pairs in an n–type semiconductor is

$$\tau = \tau_{p0} \frac{n_{n0} + n_1 + \delta p}{p_{n0} + n_{n0} + \delta p} + \tau_{n0} \frac{p_{n0} + p_1 + \delta p}{p_{n0} + n_{n0} + \delta p}. \tag{7.2}$$

The values of the coefficients n_1 and p_1 are

$$n_1 = \gamma_t N_c \exp\left(-\frac{E_c - E_t}{kT}\right), \tag{7.3}$$

$$p_1 = \gamma_t N_v \exp\left(-\frac{E_t - E_v}{kT}\right), \tag{7.4}$$

where E_t is the energy level of traps (recombination centers); γ_t is the degeneracy factor of an impurity level, i.e., the ratio of the degree of its degeneracy when the level is free and when it is occupied by a carrier. For a semiconductor with simple singly charged centers, $\gamma_t = 1$. If we assume that the trap levels are located in the upper half of the forbidden band and the semiconductor is of the n-type, the physical meaning of the coefficient n_1 (compared with the value n_{n0}) is the degree of occupancy of traps by majority carriers. When $n_1/n_{n0} \gg 1$, practically all the capture centers are empty, but when $n_1/n_{n0} \ll 1$, paractically all the centers are occupied. The coefficient p_1 has a similar meaning when the trap level is in the lower half of the forbidden band of a p–type semiconductor.

The coefficient τ_{p0} in Eq. (7.2) represents the lifetime of holes in a heavily doped n–type semiconductor, while τ_{n0} represents the electron lifetime in a heavily doped p–type semiconductor.

The formulas for these lifetimes can be obtained easily from simple physical considerations. In heavily doped semiconductors, all traps are filled with electrons and therefore each hole capture results in the recombination of a carrier pair. Since a trap is equivalent to a circle of area σ_p, the probability of capture of a hole moving at a thermal velocity v_p is $N_t \sigma_p v_p$ (here, N_t is the concentration of traps). The hole lifetime is simply the reciprocal of the recombination probability. Therefore,

$$\tau_{p0} = \frac{1}{N_t\,\sigma_p\,v_p} \tag{7.5}$$

is the hole lifetime in a heavily doped n-type semiconductor and, correspondingly, the electron lifetime in a heavily doped p-type semiconductor is

$$\tau_{n0} = \frac{1}{N_t\,\sigma_n\,v_n}. \tag{7.6}$$

The expression (7.2) explains correctly the experimentally observed dependence of the lifetime on the Fermi level position, on the temperature, and on the injection level. At low and high injection levels, the general expression for the lifetime can be simplified to:

$$\tau_p \simeq \tau_{p0}\frac{n_{n0}+n_1}{n_{n0}+p_{n0}} + \tau_{n0}\frac{p_{n0}+p_1}{n_{n0}+p_{n0}} = \tau_0 \qquad (\delta p \ll n_{n0}), \tag{7.7}$$

$$\tau_p \simeq \tau_{p0} + \tau_{n0} = \tau_\infty \qquad\qquad (\delta p \gg n_{n0}). \tag{7.8}$$

At moderate values of the injection level $(\Delta = \delta_p/n_{n0})$, the lifetime is not constant and, in the case $n_{n0} \ll p_{n0}$ (an n-type semiconductor far from intrinsic conduction conditions), the lifetime is given by

$$\tau(\Delta) \simeq \frac{1+\frac{\tau_\infty}{\tau_0}\Delta}{1+\Delta}\,\tau_0. \tag{7.9}$$

Thus, when $\Delta \ll 1$ and $\Delta \gg 1$, the hole lifetime is τ_p = const, and this assumption has been made in our discussion of transient processes in diodes.

However, real recombination conditions in semiconductor crystals are considerably more complex than those assumed in the simple Shockley-Read model. First of all, a semiconductor usually contains several types of recombination center and each type may have several energy levels in the forbidden band. Moreover, during the switching of a diode, recombination takes place under non-stationary conditions and the degree of departure from thermodynamic equilibrium varies considerably over the volume of the diode.

For these reasons, it is important to find to what degree the recombination processes in the base of a real semiconductor diode approach, during switching, the monomolecular recombination model in which the lifetime has a constant value independent of the excess carrier density.

§ 25. LIFETIME OF HOLES UNDER VARIOUS RECOMBINATION CONDITIONS

25.1. Lifetime Under Nonstationary Conditions

To determine the time dependence of the nonequilibrium carrier density under nonstationary conditions in the case of nonuniform excitation (this corresponds to the switching of a diode from the forward to the reverse direction), it is necessary to solve a system of transport equations for electron transitions between allowed energy levels. In the general case of an arbitrary departure from equilibrium and several energy levels of a trap (or several simple traps), the dependences $\delta p(t)$ and $\delta n(t)$ are extremely complex. Even in the case of a semiconductor with one type of singly charged trap, the lifetime under stationary conditions in the case of an arbitrary injection level cannot be found analytically because the kinetics of the rise or decay of the excess carrier density is described by nonlinear differential equations [127].

In the case of small departures from equilibrium ($\Delta \ll 1$), we can neglect terms which are quadratic in respect of the nonequilibrium density and the corresponding time dependences can be made linear. Then, the solution for $\delta n(t)$ and $\delta p(t)$ can be represented by the sum of exponential terms of the type

$$\delta p(t) = \Sigma A_i e^{-t/\tau_i}. \tag{7.10}$$

The recombination statistics for the nonstationary case in a semiconductor with one type of simple recombination center has been considered by Adirovich and Guro [128, 141].

The analysis given in [128] shows that in an n-type semiconductor the decay of the nonequilibrium hole density after the end of excitation is described by two exponential terms and depends on three time constants, one of which is the steady-state lifetime of holes τ_p^{sts} obtained by Shockley and Read in the case of arbitrary concentration of recombination centers (traps) N_t:

$$\tau_p^{sts} = \frac{\tau_{n0}(p_{n0} + p_1) + \tau_{p0}\left[n_{n0} + n_1 + N_t\left(1 + \dfrac{n_{n0}}{n_1}\right)^{-1}\right]}{n_{n0} + p_{n0} + N_t\left(1 + \dfrac{n_{n0}}{n_1}\right)^{-1}\left(1 + \dfrac{n_1}{n_{n0}}\right)^{-1}}. \tag{7.11}$$

The other two time constants, τ_1 and τ_2, which occur in the exponential functions are themselves functions of the parameters τ_{p0}, τ_{n0}, n_{n0}, p_{n0}, n_1, p_1, and of the degree of occupancy of traps in the equilibrium state. When the concentration of traps is low ($N_t \ll n_{n0}$), it is found that $\tau_2 \approx \tau_p^{sts} = \tau_0$ and $\tau_1 \ll \tau_0$, i.e., the relaxation processes are described by a single time constant, which is equal to the hole lifetime given by Eq. (7.7). When the concentration of traps is very high ($N_t \gg n_{n0}$), the decay of the nonequilibrium hole density is also described by a single time constant but, in this case, this constant is equal to the lifetime of the majority carriers, i.e., electrons.

Similar expressions have been obtained for τ_1 and τ_2 by Sandiford [129]. Using typical values of the capture cross sections and energy levels of traps in silicon and germanium, Sandiford has established that we always have $\tau_2 \gg \tau_1$, so that − with the exception of a short initial time interval − the recombination decay of the excess hole density is described by the time constant τ_2. After experimentally justified simplifications, Sandiford has obtained a convenient formula for analysis of the nonstationary hole lifetime in an n-type semiconductor with an arbitrary trap concentration provided the injection level is low:

$$\tau_p^{nst} = \frac{\tau_{n0}\left[p_{n0} + p_1 + N_t\left(1 + \frac{n_1}{n_0}\right)^{-1}\right] + \tau_{p0}\left[n_{n0} + n_1 + N_t\left(1 + \frac{n_0}{n_1}\right)^{-1}\right]}{n_{n0} + p_{n0} + N_t\left(1 + \frac{n_1}{n_{n0}}\right)^{-1}\left(1 + \frac{n_{n0}}{n_1}\right)^{-1}} \tag{7.12}$$

Comparison of Eqs. (7.11) and (7.12) shows that the expression for the steady-state lifetime includes N_t only in the term describing the capture of the minority carriers by traps, while the formula for τ_p^{nst} is symmetrical with respect to N_t ("nst" denotes "nonstationary state").

In the majority of cases of practical interest, Eqs. (7.11) and (7.12) can be simplified considerably.

Let us consider an n-type semiconductor under conditions ($n_{n0} \gg p_{n0}$) with recombination centers having an energy level in the upper half of the forbidden band so that $n_1 \gg p_1$. We shall also assume that the quantities τ_{p0} and τ_{n0} do not differ greatly from one another. The results of the simplification of Eqs. (7.7), (7.11), and (7.12) for various relationships between n_{n0}, n_1, and N_t are given in Table 7.1.

The case $n_{n0} \gg n_1$ closely approximates to gold–doped n–type silicon, while the cases $n_{n0} \ll n_1$ and $n_{n0} = n_1$ correspond to gold-doped n–type germanium (cf. § 27). We must point out that very high trap concentrations ($N_t \gg n_{n0}$) are rarely encountered in practical diodes. This is because those impurities which strongly reduce the carrier lifetime in germanium and silicon act also as donors or acceptors; therefore, to avoid the dependence of the resistivity of a crystal on the concentration centers it is usual to employ only those samples in which $N_t < n_{n0}$. We can arbitrarily assume that the upper limit of N_t, which we may encounter in practical diodes, is given by the equality $N_t = n_{n0}$.

Analysis of the formulas in Table 7.1 yields the following conclusions. First of all, we find that for any trap concentration the steady–state hole lifetime is equal to τ_0, which is given by the Shockley–Read statistics for $N_t \ll n_{n0}$. At low values of N_t the nonstationary hole lifetime is also equal to τ_0. When the trap concentration is increased, the value of τ_p^{nst} begins to be governed more and more by the value of τ_{n0}, which is the majority–carrier lifetime. However, we always have $\tau_p^{nst} > \tau_p^{sts}$.

In considering transient processes in diodes, we need not always use the value of τ_p^{nst}. Thus, in the determination of the first (recovery) phase duration t_1 under switching conditions, when $t_1 \ll \tau_p$ (i.e., when $B \geq 1$), we can use the steady–state hole lifetime corresponding to the given conditions (n_{n0}, n_1, N_t), because during the time t_1 the degree of carrier occupancy of traps does not change appreciably. This confirms the validity of calculations of the recovery phase duration t_1 [44, 45] based on the use of an expression for the steady–state hole lifetime.

In the investigation of the second phase of the transient reverse current decay or the postinjection voltage decay across the p–n junction, we must use the value of τ_p^{nst}.

The general conclusion of this analysis is that, in all cases of low injection levels, the lifetime of excess holes is independent of the hole density. Transient processes in diodes of duration less than τ_p are governed by the steady–state lifetime, which is identical with the hole lifetime given by Eq. (7.7); in considering longer

Table 7.1

n_{n0}	N_t	τ_0	τ_p^{sts}	τ_p^{nst}
$n_{n0} > n_1$	$N_t \ll n_{n0}$	$\tau_{p0}\left(1+\dfrac{n_1}{n_{n0}}\right) \simeq \tau_{p0}$	$\tau_{p0}\left(1+\dfrac{n_1}{n_{n0}}\right) \simeq \tau_{p0}$	$\tau_{p0}\left(1+\dfrac{n_1}{n_{n0}}\right) \simeq \tau_{p0}$
	$N_t = n_{n0}$	—	τ_{p0}	$\tau_{p0} + \tau_{n0}$
	$N_t \gg n_{n0}$	—	τ_{p0}	$\tau_{p0} + \tau_{n0}\dfrac{n_{n0}}{n_1}$
$n_{n0} = n_1$	$N_t \ll n_{n0}$	$2\tau_{p0}$	$2\tau_{p0}$	$2\tau_{p0}$
	$N_t = n_{n0}$	—	$2\tau_{p0}$	$2\tau_{p0} + 0.4\tau_{n0}$
	$N_t \gg n_{n0}$	—	$2\tau_{p0}$	$2\left(\tau_{p0} + \tau_{n0}\right)$
$n_{n0} < n_1$	$N_t \ll n_{n0}$	$\tau_{p0}\left(1+\dfrac{n_1}{n_{n0}}\right) \simeq \tau_{p0}\dfrac{n_1}{n_{n0}}$	$\tau_{p0}\left(1+\dfrac{n_1}{n_{n0}}\right) \simeq \tau_{p0}\dfrac{n_1}{n_{n0}}$	$\tau_{p0}\left(1+\dfrac{n_1}{n_{n0}}\right) \simeq \tau_{p0}\dfrac{n_1}{n_{n0}}$
	$N_t = n_{n0}$	—	$\tau_{p0}\left(2+\dfrac{n_1}{n_{n0}}\right) \simeq \tau_{p0}\dfrac{n_1}{n_{n0}}$	$\tau_{p0}\left(2+\dfrac{n_1}{n_{n0}}\right) + \dfrac{n_{n0}}{n_1}\tau_{n0} \simeq \tau_{p0}\dfrac{n_1}{n_{n0}}$
	$N_t \gg n_{n0}$	—	$\tau_{p0}\left(1+\dfrac{n_1}{n_{n0}}\right) \simeq \tau_{p0}\dfrac{n_1}{n_{n0}}$	$\tau_{p0}\left(1+\dfrac{n_1}{n_{n0}}\right) + \tau_{n0} \simeq \tau_{n0}\dfrac{n_1}{n_{n0}}$

transient processes, we must use the nonstationary value of τ_p, which is approximately equal to the sum of the steady-state hole and electron lifetimes.

25.2. Recombination at Multiply
Charged Centers

It also follows from our analysis that during transient pro-cessses in diodes the most active are those recombination centers which have high electron and hole capture probabilities, i.e., those whose energy levels lie close to the middle of the forbidden band (they are known as deep levels).

Experimental investigations have demonstrated that such re-combination centers are formed in silicon and germanium by atoms of transition metals as well as of copper and zinc. It has been es-tablished that, in contrast to simple centers which have one energy level in the forbidden band, atoms of these metals have a complex energy spectrum in semiconductors with several levels in the for-bidden band. Since such "complex traps" can have several charged states, they are known as multiply charged centers.

Because of a strong interaction between charges of impurity centers, it is found that not all the possible levels are observed experimentally but only one or two [130].

Analysis of the recombination processes involving multiply charged centers is very difficult. This difficulty is due to the fact that, in spite of the constancy of the concentration of such centers, the density of energy levels corresponding to differently charged atoms is not constant because it depends on the degree of occupancy of these centers with electrons and this degree varies during the recombination process itself.

However, a mathematical analysis [130] shows that if the sep-ations between energy levels are considerably greater than kT/q (which is true of the majority of investigated impurities), the dependence of the effective charge of the recombination centers on the Fermi level is a step function, i.e., recombination takes place only at one or two levels.

Therefore, to determine the recombination properties of real crystals, it is sufficient to consider a semiconductor with traps of one kind having two levels in the forbidden band at E_1 and E_2. This problem has been considered by Shockley and Sah [132], as well as by Zhdanova, Kalashnikov, and Morozov [133]. Using a general expression for steady-state rate of generation of carriers at multiply charged centers [132], Kontsevoi [131] has derived formula for the hole lifetime (in an n-type semiconductor) for an arbitrary injection level:

$$\tau_p = \tau_0^{(2)} \frac{1 + a\Delta + b\Delta^2}{(1 + \Delta)(1 + c\Delta)}, \qquad (7.13)$$

where

$$a = \frac{(c_{n2} + c_{p2})\left(c_{n1} + c_{p1}\frac{p_1}{n_{n0}}\right) + c_{n2}\left[c_{n1}\left(1 + \frac{n_2}{n_{n0}}\right) + c_{p1}\frac{n_2}{n_{n0}}\right]}{c_{n2}\left(c_{n1} + c_{p1}\frac{p_1}{n_{n0}}\right)\left(1 + \frac{n_2}{n_{n0}}\right)}, \qquad (7.14)$$

$$b = \frac{c_{n1}c_{n2} + c_{p1}c_{p2} + c_{n1}c_{p2}}{c_{n2}\left(c_{n1} + c_{p1}\frac{p_1}{n_{n0}}\right)\left(1 + \frac{n_2}{n_{n0}}\right)}, \qquad (7.15)$$

$$c = \frac{c_{n1}c_{p2}(c_{p1} + c_{n2})}{c_{n1}c_{n2}c_{p2} + c_{p1}c_{n2}\left(c_{n1}\frac{n_2}{n_{n0}} + c_{p2}\frac{p_1}{n_{n0}}\right)}. \qquad (7.16)$$

The coefficients $c_{n1}, \ldots, c_{p2}$ are reciprocals of the electron and hole lifetimes in a heavily doped semiconductor with a single recombination level, i.e.,

$$c_{n1} = \frac{1}{\sigma_{n1}N_t v_n} = \tau_{n01}^{-1}. \qquad (7.17)$$

The coefficients c_{p1}, c_{n2}, and c_{p2} are given by similar expressions. The characteristic quantities p_1 and n_2 for each of the levels are determined in exactly the same manner as for a semiconductor with singly charged centers, i.e., using Eqs. (7.3) and (7.4). The subscript "1" in Eqs. (7.14)-(7.17) refers to the first level and the subscript "2" refers to the second level; the levels are numbered starting from the top of the valence band E_V.

In the derivation of Eq. (7.13), we have assumed that the levels 1 and 2 lie on the opposite sides of the middle of the forbidden band and sufficiently far from it so that $p_1 \gg n_1$ and $n_2 \gg p_2$. Therefore, the formula for τ_p includes only the coefficients in front of p_1 and n_2.

The value of the hole lifetime at low injection levels is

$$\tau_0^{(2)} = \tau_{p02}\left(1 + \frac{n_2}{n_{n0}}\right)\left(1 + \frac{1}{\dfrac{c_{p2}}{c_{p1}}\dfrac{n_{n0}}{n_2} + \dfrac{c_{p2}}{c_{n1}}\dfrac{p_1}{n_2}}\right)^{-1}.$$

(7.18)

If all the coefficients $c_{n_1}, \ldots, c_{p_2}$ have similar values and the second level is considerably deeper thant the first (i.e., when $n_2 \ll p_1$), it follows that $\tau_0^{(2)} \simeq \tau_{p02}$, i.e., we have the same situation as in the case of a simple singly charged center with a level at E_2. We obtain the same result also when the depth of both levels is the same $(p_1 \approx n_2)$ but $c_{p2} \gg c_{n1}$.

At very high injection levels $(\Delta \to \infty)$ the hole lifetime is

$$\tau_\infty^{(2)} = \tau_0^{(2)}\frac{b}{c} = \frac{c_{n1}c_{n2} + (c_{n1} + c_{p1})\,c_{p2}}{c_{n1}c_{p2}\,(c_{p1} + c_{n1})}$$

(7.19)

and it depends only on the electron and hole capture cross sections of the levels E_1 and E_2 (as in the case of recombination at a singly charged center). If the capture cross sections of one of the levels are considerably larger than those of the other level, then

$$\tau_\infty^{(2)} = \tau_{p02} + \tau_{n02}\left(1 + \frac{c_{p1}}{c_{p2}}\right) \qquad (c_{n,\,p1} \ll c_{n,\,p2}),$$

(7.20)

$$\tau_\infty^{(2)} = \tau_{p01}\left(1 + \frac{c_{n2}}{c_{p2}}\right) + \tau_{n01} \qquad (c_{n,\,p1} \gg c_{n,\,p2}),$$

(7.21)

i.e., as before, the hole lifetime is governed by the parameters of both levels of the recombination centers.

This analysis shows that in the case of recombination at a multiply charged trap with two levels in the forbidden band, as well as in the case of simple centers, the hole lifetime for $\Delta \ll 1$ and

$\Delta \gg 1$ is independent of the nonequilibrium carrier density. When the injection level increases from $\Delta = 0$ to $\Delta \to \infty$, the lifetime varies monotonically from $\tau_p = \tau_0^{(2)}$ to $\tau_p = \tau_\infty^2$.

Thus, in the majority of cases of recombination at traps, we can still use the parameter τ_p (which is independent of the density of nonequilibrium carriers) to analyze transient processes in diodes. However, the value of τ_p may depend on the concentration and properties of recombination centers in the base and on the nature of the nonstationary process being considered.

25.3. Radiative Recombination

At very high injection levels, when the density of excess carriers in the base is at least 10^{18} cm^{-3}, recombination begins to obey the dimolecular law, i.e., the hole lifetime is inversely proportional to the density of holes. This is because, in addition to the trapping mechanism, direct radiative recombination of free electrons and holes becomes possible.

Analysis of radiative recombination, i.e., recombination resulting in the emission of photons, was carried out by Shockley and van Roosbroeck [134]. This analysis yields the following relationship for the excess carrier lifetime:

$$\tau_{rad} \simeq \frac{n+p}{np} R_c. \tag{7.22}$$

In this expression τ_{rad} represents the lifetime of carriers which would apply if only the radiative recombination mechanism were active. The coefficient R_c has been determined experimentally for germanium [134] and at room temperature it is $1.57 \cdot 10^{13}$ cm$^{-3} \cdot$ sec^{-1}

It follows from Eq. (7.22) that, when $n = p = 10^{18}$ cm^{-3}, the radiative lifetime of holes is $\tau_{rad} \approx 30$ μsec, which is comparable with the "trapping" lifetime observed in some diodes. When the temperature is increased, the role of radiative recombination increases rapidly. Thus, at high injected carrier densities and at elevated temperatures the constancy of τ_p may not be observed and a consequence of this would be a deviation from the basic linear equation (1.18).

§ 26. INFLUENCE OF TRAPPING LEVELS ON TRANSIENT PROCESSES IN DIODES

26.1. Trapping Levels

In the preceding section, we have pointed out that the larger the carrier capture cross section of a center, the more effective is the recombination through this center.

However, the probability of capture of an electron or hole does not yield directly the value of the recombination probability. An electron captured by a vacant center can either remain at this center (until a free hole is captured with which the electron recombines), or this electron may be transferred back to the conduction band as a result of thermal fluctuations. The relative importance of these two processes is governed by the positions of the energy levels of the recombination centers in the forbidden band, the concentration of such centers, and the free carrier density, i.e., the Fermi level position.

If, for example, the trapping level is very close to the lower edge of the conduction band, the probability of capture of an electron by a trap is high. On the other hand, the probability of capture of a hole by such a center is almost the same as the probability of a transition of a hole through the whole forbidden band, i.e., it is practically equal to zero. Thus, electrons captured by such a trap do not have a chance to recombine with holes and they return to the conduction band by thermal processes.

The same applies to holes trapped at centers located immediately above the top of the valence band. Centers for which the probability of a return transition of a carrier to the conduction or valence band is considerably higher than the probability of the capture of a carrier of opposite sign are known as trapping centers or levels. In other words, a trapping level interacts effectively with only one type of carrier. Trapping centers, like recombination centers, are present in every real semiconductor crystal.

Experiments show that a carrier captured by a trapping level can sometimes reside for a very long time ($\tau_t \gg \tau_p$) in the localized state and then may be liberated again.

Trapping levels can be divided into two types in accordance with their interaction with the conduction band. When the establishment of equilibrium between the trapping levels and the conduction band occurs in a time much shorter than τ_p, we speak of α-type centers. A carrier captured by an α-type center may be liberated and captured several times before it recombines with a carrier of opposite sign. Therefore, α-type centers are known as multiple trapping levels.

When the establishment of equilibrium between the conduction band and the trapping levels takes a time much longer than the carrier lifetime, such trapping levels are known as β-type. These β-type centers are known as single trapping levels.

Experiments of Hornbeck and Haynes [135] have shown that typical values of the trapping time τ_t for silicon are 50 msec for α-type centers and more than 1 sec for β-type centers. The values of the concentrations of the α- and β-type centers, N_α and N_β, are usually 10^{13}–10^{14} cm^{-3} or less.

Trapping levels for which $\tau_t \approx 1$ msec are also observed in germanium at low temperature [136].

26.2. Role of Trapping Levels in
Devices with p-n Junctions

Under steady-state conditions, during prolonged flow of a constant forward current through a diode, we have – apart from the charge of mobile holes stored in the base – an additional charge of holes localized at trapping levels. Thus, the presence of trapping levels prolongs the relaxation time of the nonequilibrium electron and hole densities and, therefore, we can expect a corresponding slowing down of transient processes during the switching of a diode.

However, experimental investigations, carried out by Ryvkin and his colleagues on photoelectric devices with p-n junctions [137, 138], have demonstrated that trapping levels alter appreciably the photoconductivity decay curves but do not affect greatly the photocurrent relaxation.

Analysis of extensive experimental data on transient processes in various types of diode also shows that the role of trapping levels

is slight. Comparison of transient switching characteristics of alloyed silicon diodes, in which the value of τ_p in the base has been
reduced to 0.1-0.2 μsec by various methods (thermal quenching,
neutron bombardment, doping with gold), has shown that the form
of the function i(t) does not differ greatly for diodes subjected to
these three treatments. The reverse current decay right down to
5-10 μA is described satisfactorily by Eq. (2.4) with a constant value of τ_p. The cause of the different effect of trapping levels in
different transient processes is as follows [138]. That fraction of
holes injected into the base (irrespective of whether they are injected optically or by a p-n junction biased in the forward direction)
which is localized at trapping centers becomes liberated and is
responsible for the slow component of the transient current. However, because of the extremely small number of carriers liberated
thermally from trapping centers, this slow component of the current
is of very low amplitude compared with the current due to the relaxation of free excess carriers. Therefore, the liberation of carriers from trapping levels has practically no influence on the shape
of the oscillograms of the transient processes in diodes and photodiodes.

This is particularly evident in the presence of β-type centers
because of the very large values of τ_t characterizing these centers.
In the case of α-type centers, the influence of trapping on the rate
of relaxation of the current in fast photodiodes (with a response
time $\sim 10^{-8}$ sec) may become appreciable only for $N_\alpha \gtrsim 10^{15}$ cm^{-3}
[138]. In the estimate of the influence of these α-type centers, we
have assumed that these trapping levels have the largest known
carrier capture cross sections. In photodiodes with a slower response, the α-type centers may affect the response at lower values
of N_α.

The estimates reported in [138] were made on the assumption of uniform carrier generation throughout the base, which does
not agree with the conditions during the flow of the forward and reverse currents through the p-n junction.

Adirovich et al. [139] developed a diffusion theory of p-n
junctions for semiconductors with trapping centers. They established that trapping processes do not affect the steady-state distribution of free holes in the base during the flow of the forward
current. This is due to the self-evident circumstance that, under

steady-state conditions, the capture of nonequilibrium carriers at trapping levels is balanced at each point in the semiconductor by the liberation of carriers from these levels. In view of the presence of a density gradient of the generated carriers, the diffusion of carriers takes place irrespective of whether the trapping is active. Lashkarev established theoretically [140] and Hornbeck and Haynes confirmed experimentally [135] that the diffusion length is practically unaffected by the presence of trapping centers.

Under transient conditions, the rate of capture of holes by trapping levels is not equal to the rate of liberation. Therefore, the differential impedance of a p-n junction for a small high-freqency signal can sometimes depend not only on the hole lifetime τ_p but also on the parameters of the trapping centers (τ_t, N_α, E_α, or N_β, E_β [139].

26.3. Trapping Levels During
Switching of a Diode

Let us now compare the charge of free and localized holes stored in the base during the flow of the forward current. The charge of free holes has been derived earlier and, according to Eq. (2.8), is given by $Q_{st} = i_f \tau_p$.

Assuming that during the flow of the forward current all the trapping levels are saturated with holes and that, after the switching of the diode, only those carriers take part in the reverse current relaxation which are within the diffusion length from the p-n junction, we obtain the charge localized at trapping levels

$$Q_{loc} \simeq qSL_p N_t. \tag{7.23}$$

Comparison of Q_{loc} with Q_{st} shows that, in the case of low values of the forward current and a large-area p-n junction (i.e., when the forward current density is low), the ratio Q_{loc}/Q_{st} may be very large.

However, in practice, the forward current density during the switching processes is made reasonably high in order to reduce the influence of the charge of the p-n junction barrier capacitance

Q_C. We shall assume arbitarily that the minimum forward current pulse is set by the condition $Q_C < 0.1 \, Q_{st}$. For sufficiently high reverse voltages ($u_r \gg u_f$) in the case of a p-n junction with an abrupt impurity gradient, we can easily obtain the following expression, using the well-known formula for the barrier capacitance:

$$Q_c \simeq \frac{KS}{\sqrt{\rho}} \sqrt{u_r} \, , \tag{7.24}$$

where $K = 4.6 \cdot 10^{-8} \text{sec} \cdot \text{cm}^{-1/2} \cdot \text{A}^{1/2}$ for germanium and silicon (to within 15%). Hence,

$$i_f^{(\min)} \gtrsim \frac{10 K S \sqrt{u_r}}{\tau_p \sqrt{\rho}} \tag{7.25}$$

Comparison of the expressions for Q_{st} and Q_{loc} using Eq. (7.25) gives

$$Q_{loc} \leqslant Q_{st} \frac{N_t L_p q \sqrt{\rho}}{10 \, K \sqrt{u_r}} . \tag{7.26}$$

Using the least favorable combination of the parameters ($L_p \approx 10^{-2}$ cm, $\rho \approx 1 \, \Omega \cdot$ cm, $u_r \approx 4$ V), we find that Q_{loc} is comparable with Q_{st} only when $N_t > 6 \cdot 10^{14}$ cm^{-3}, which is close to the maximum trap concentration in real crystals. Since the time of dispersal of the free hole charge is at least two or three orders of magnitude shorter than the time of dispersal of the charge localized at trapping levels, it follows that the amplitude of the signal associated with the liberation of carriers from trapping levels is two or three orders of magnitude smaller than the amplitude of the current observed on the screen of an oscillograph during the switching of a diode.

Obviously, the trapping centers of the α-type may affect considerably the transient process in the case of short forward current pulses ($t_f \ll \tau_p$). This problem has not been considered specially in the literature.

§27. RECOMBINATION PROPERTIES OF GOLD-DOPED GERMANIUM AND SILICON

27.1. General Properties of Multiply Charged Centers

We have mentioned in §25 that centers formed by transition metal atoms as well as by copper and zinc are most effective in recombination processes in germanium and silicon.

Figure 7.1 shows the experimental energy spectra of multiply charged centers in germanium and silicon. The energy of each level is measured either from the top of the valence band E_V or the bottom of the conduction band E_C, depending whether the level is closer to the valence or conduction band. Table 7.2 summarizes the numerous experimental data on the electron (σ_n) and hole (σ_p) capture cross sections of atoms of various elements, in various charge states, present in germanium. For the sake of comparison, we can mention that simple singly charged centers, such as those formed by gallium or aluminum atoms, have electron capture cross sections not larger than 10^{-18}-10^{-19} cm^2.

Atoms of transition metals, as well as copper and zinc atoms, present in germanium and silicon have certain properties in common and knowledge of these properties is important in the analysis of recombination processes.

1. Transition metals have higher diffusion coefficients in solid semiconductors and lower segregation coefficients as well as maximum solubilities lower than elements of groups III and V in the periodic table.

2. The recombination centers formed by transition metal atoms have large electron and hole capture cross sections.

3. The majority of transition metals (Ni, Fe, Cu, Co in germanium, and Fe, Cu in silicon) may suffer "deactivation", i.e., they may lose recombination properties after prolonged annealing. The "deactivation effect" is due to the high values of the diffusion coefficients and the strong temperature dependences of the solubility of these impurities which result in the "precipitation" of the atoms of

these metals at interstices in the crystal lattice or in the coagulation of these atoms during annealing; this destroys the recombination properties of these atoms.

4. The carrier capture cross sections of transition metal atoms depend strongly on temperature. When the temperature is increased from room temperature to 120-150°C, the capture cross sections decrease by a factor of 2-7, depending on the impurity.

The third of the properties just listed has resulted in the almost exclusive use of gold in the reduction of the carrier lifetime in materials used to make fast-response silicon and germanium diodes.

27.2. Gold-Doped Germanium

Atoms of gold in germanium may have five charged states: they may be neutral atoms, singly charged positive ions, or singly,

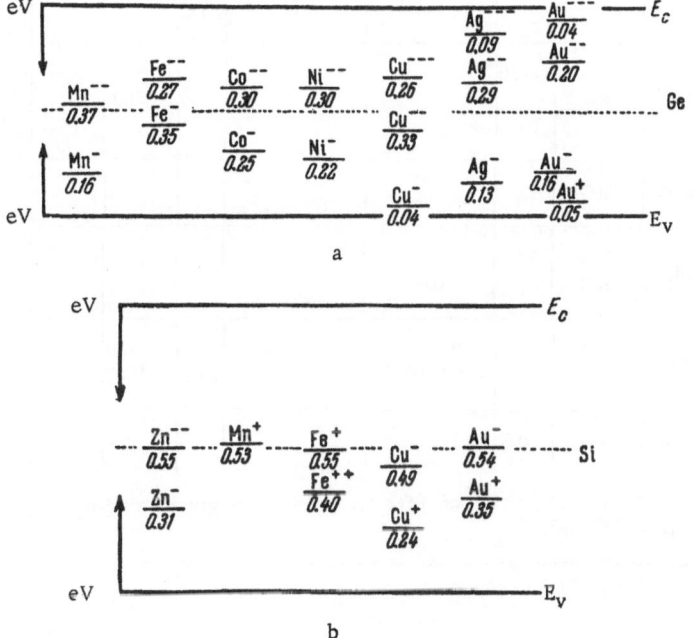

Fig. 7.1. Energy spectrum of multiply charge centers in germanium(a) and silicon (b).

Table 7.2

Element	T, °K	σ_n^0	σ_n^-	σ_n^{--}	σ_p^-	σ_p^{--}	σ_p^{---}
Cu	300 316 130—190 20		0.1 0.15 0.18	0.36 $\leqslant 0.001\sigma_p^{---}$ $\sim 0.01\sigma_p^{---}$	500	1.8 $\sim 10\sigma_n^-$	1 1 3.6
Ni	300 77	0.8 1 20 7.8	6 3		1000	> 40 100 200	
Fe	300	1—10 10			30	100	
Co	300	10					
Mn	300 150	30	< 0.0001				
Au	300 80	1.2 600	2 ~ 0.1 (90° K)		1000	100 > 1000	
Ag	300	0.6	1			50	

Note. The value of σ for all elements is given in units of 10^{-16} cm².

Table 7.3

σ_n^0	σ_n^-	σ_p^0	σ_p^-	σ_p^{--}	Reference
600 (80° K)	0.1 (90° K)		1 000 (80° K)	1 000 (80° K)	[143]
1.2	2			100	[144]
		0.0375 (97° K)	230 (77° K)		[145]
			15 500 (97° K)		[146]
			4 900 (80° K)		[147]
					[146, 148, 150]
0.5	2	300 (70° K)		100	[149]
		0.006 (25° K)			[155]
	1 (300° K)			1 500	[152]
	0.12 (77° K)				[153]
			100—200 (24° K)	4 000 (30° K)	[154]

Note. The value of σ is given in units of 10^{-16} cm^2. The results for which the temperature is not quoted were obtained at room temperature.

doubly, and triply charge negative ions. The experimental values of the carrier capture cross sections of gold atoms in the various charged states are listed in Table 7.3.

Analysis of the recombination process at gold atoms in germanium is very complex if all five possible charged states are allowed for; such an analysis has not yet been carried out.

If germanium has p-type conduction and its Fermi level lies above E_2 (which is the Au⁻ level in Fig. 7.1), or if germanium has n-type conduction and the Fermi level lies below E_3 (which is the Au⁻⁻ level in Fig. 7.1), only one or two deep levels take part in the recombination process at low injection levels.

Analysis of Eq. (7.18) shows that the room-temperature electron lifetime in p-type germanium at low injection levels is given by

$$\tau_0 = \frac{p_{p0} + p_2}{p_{p0}c_{n2} + p_2 c_{n3}}. \tag{7.27}$$

In n-type germanium, the recombination properties are governed by the hole capture cross section of the level E_3 and the recombination takes place as if only a simple singly charged center were present. If $n_{n0} \gg p_{n0}$, then, according to the Shockley-Read theory, the excess hole lifetime is given by

$$\tau_0 = \frac{1}{c_{p3}}\left(1 + \frac{n_3}{n_{n0}}\right). \tag{7.28}$$

In these expressions, the subscripts "2" and "3" represent properties of the levels E_2 and E_3, respectively.

To use Eqs. (7.27)–(7.28) it is necessary to know [cf. Eqs. (7.3) and (7.4)] the degeneracy factors of the varous levels of gold. The published data on the degeneracy factors of gold are contradictory. Thus, a study of the temperature dependence of the Hall coefficient has indicated [155] that in p-type germanium the reciprocals of the degeneracy factors of all the deep acceptor levels (including the level E_3) are equal to 8, while for n-type germanium they are equal to 2. Measurements of the carrier lifetime in n-type germanium, reported in [148], have yielded $\gamma_3^{-1} = 2.6$ and $\gamma_2^{-1} = 8.5$. Since the experimental conditions used in the latter study [148] were closer to those in the processes of interest to us, we shall assume that $\gamma_3^{-1} = 2.6$ and $\gamma_2^{-1} = 8.5$. Assuming that

$$N_c = 4.82 \cdot 10^{15} T^{3/2},$$ (7.29)

we obtain the following values at room temperature (T = 300°K):

$$n_3 = 7.2 \cdot 10^{15} \text{ cm}^{-3}, \qquad p_2 = 8.8 \cdot 10^{15} \text{ cm}^{-3}.$$

The results given in Table 7.3 and the temperature dependences of the electron and hole capture cross sections of the various levels [148] show that the following values of the recombination parameters are the most reliable in the case of the levels E_2 and E_3:

$$\sigma_{p2}^{-} = 3 \cdot 10^{-14} \text{ cm}^2, \qquad \sigma_{p3}^{--} = 10^{-14} \text{ cm}^2,$$
$$\sigma_{n2}^{-} = 5 \cdot 10^{-17} \text{ cm}^2, \qquad \sigma_{n3}^{--} = 2 \cdot 10^{-16} \text{ cm}^2.$$

In the determination of these cross sections, we have arbitrarily assumed that $v_p = v_n = 10^7$ cm/sec.

When the electrical conductivity of a crystal is moderate and still far from intrinsic conditions, the value of the conductivity can be assumed to be directly proportional to the majority carrier density. In this case, simple transformations to Eqs. (7.27)–(7.28) yield the following expressions:*

$$\tau_0 = \frac{2 \cdot 10^9}{N_{Au}} \cdot \frac{1 + 3\rho}{1 + 12\rho} \quad \text{for p-type germanium,}$$ (7.30)

$$\tau_0 = \frac{10^7}{N_{Au}} (1 + 3.6\rho) \quad \text{for n-type germanium.}$$ (7.31)

In these formulas, the value of N_{Au} is expressed in cm^{-3}, where ρ is in $\Omega \cdot \text{cm}$ and τ_0 is in seconds.

---

* The numerical factors in these relationships are, as is clear from their derivation, somewhat arbitrary.

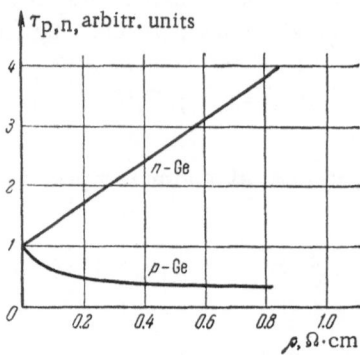

Fig. 7.2. Dependence of the minority-carrier lifetime in gold-doped germanium on the electrical resistivity.

Naturally, for p- and n-type germanium we find that $\tau_0 \sim N_{Au}^{-1}$. However, the dependence of τ_0 on the resistivity may assume various forms. Figure 7.2 shows graphically the dependences (7.30) and (7.31) in relative units. We can see that as ρ_n increases, the value of τ_0 increases linearly, but when ρ_p increases the electron lifetime decreases, tending to a limit which is about 25% of the value for $\rho_p \rightarrow 0$. The physical meaning of this result is that when the electrical resistivity of p-type germanium is increased, the Fermi level rises at an increasing rate toward the middle of the forbidden band. The level E_2 is then filled with electrons, practically all the gold atoms become negative and singly charged, and the recombination velocity is governed by the value of τ_{n03}, while in the case of the Fermi level lying close to E_V the recombination velocity is governed by the larger of the lifetimes τ_{n02} and τ_{n03}, i.e., by τ_{n02}. The ratio τ_{n02}/τ_{n03} is equal to 4.

In practice, the value of τ_0 in p-type germanium ceases to vary appreciably at resistivities $\rho_p \gtrsim 0.3\text{-}0.6 \ \Omega \cdot \text{cm}$.

Using Eqs. (7.30) and (7.31), we can estimate the minimum lifetimes in gold-doped germanium.

According to generally accepted data [156, 157], the maximum solubility of gold in solid germanium is $2 \cdot 10^{16} \ \text{cm}^{-3}$. Radioactivation analysis, carried out by the present author together with

Gubenko and Postnikova on germanium plates containing diffused gold, has shown that a more reliable value of the maximum solubility of gold is $4 \cdot 10^{16}$ cm^{-3}. This value is supported by measurements of the electrical properties of recrystallized gold-doped germanium films [158].

Assuming that $N_{Au} = 4 \cdot 10^{16}$ cm^{-3}, we find that the minimum values of τ_0 are 0.25 nsec and 12.5 nsec for n- and p-type germanium, respectively.

The quoted minimum value of τ_0 in p-type germanium has been reached in investigations of transient processes in some fast-response diffused mesa-type diodes [159, 160].

Similar investigations have been carried out by the present author, together with Postnikova, on diodes prepared from gold-doped n-type germanium. Since the duration of transient processes in these diodes is very short ($\sim 10^{-9}$ sec), it is necessary to have an exceptionally small barrier capacitance in order to observe these processes (this capacitance should be of the order of several tenths of a picofarad). In order to satisfy this requirement, p-n junctions of this type have been prepared by welding very thin gold wires to germanium crystals. The rectifying contact areas of diodes prepared in this way have not exceeded 10^{-5} cm^2. Investigations of these diodes have demonstrated that the experimentally determined dependence of the recovered charge of a diode on the resistivity of germanium and on the concentration of gold is close to the calculated dependence if it is assumed that the hole lifetime in the base is $\tau_p = (1-2)\tau_0$.

The hole lifetime in the diode base, determined from the dependence $Q_{rec} = f(i_f)$ is also close to the value of τ_0.

However, we must point out that the conditions under which the measurements have been carried out on diodes made from p- and n-type germanium represented high injection levels. Assuming that at high injection levels recombination still takes place only at the evergy levels E_2 and E_3^* and using Eq. (7.13) for p-type ger-

---

* This is valid in the case of injection levels for which the quasi-Fermi level of holes lies above E_1 and the quasi-Fermi level of electrons lies below E_4.

manium with $\rho_p = 0.5 \ \Omega \cdot cm$ and for n-type germanium with $\rho_n =$ 0.1 $\Omega \cdot cm$ (such crystals were used to prepare the investigated diodes), we obtain:*

$$\tau(\Delta) = \tau_0 \frac{1 + 37\Delta + 90\Delta^2}{(1+\Delta)(1+0.6\Delta)} \quad \text{for n-type germanium,} \tag{7.32}$$

$$\tau(\Delta) = \tau_0 \frac{1 + 1.4\Delta + 0.4\Delta^2}{(1+\Delta)(1+0.14\Delta)} \quad \text{for p-type germanium.} \tag{7.33}$$

These dependences are shown graphically in Fig. 7.3 for the case $N_{Au} = 1.3 \cdot 10^{16} \ cm^{-3}$. The lifetime at very high injection levels, τ_∞, can be seen to be the same for p- and n-type germanium, provided the concentration of gold in the two materials is the same. The theoretical relationship for p-type germanium, $\tau_\infty/\tau_0 = 3$, is not in conflict with the experimentally observed transient processes in diodes made from such germanium. The ratio $\tau_\infty/\tau_0 = 150$, which is predicted theoretically for n-type germanium, has not yet been confirmed experimentally.

---

*We must remember that Eq. (7.13) gives the steady-state hole lifetime.

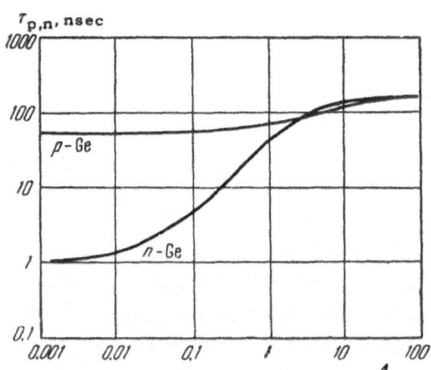

Fig. 7.3. Dependence of the minority-carrier lifetime in gold-doped germanium on the injection level Δ for ρ_p = 0.5 $\Omega \cdot cm$, ρ_n = 0.1 $\Omega \cdot cm$.

27.3. Gold-Doped Silicon

Gold atoms in silicon can have three charged states. They can be singly charged negative, singly charged positive, or neutral. Investigations of Collins, Carlson, and Gallagher [161] have demonstrated that the donor level of gold lies 0.35 eV above the top of the valence band and the acceptor level lies at a depth of 0.54 eV from the bottom of the conduction band. The electron and hole capture cross sections of these levels, obtained by various workers, are listed in Table 7.4.

Investigations of diffused silicon mesa-type diodes, prepared by the diffusion of gold at various temperatures [165], have demon-

Table 7.4

σ_n^-	σ_p^-	σ_n^+	σ_p^+	Reference
5	10	3,5 10^3 (77° K)	1	[162] [163]
$1.65 \pm 15\%$	$115 \pm 30\%$·	$63 \pm 25\%$	$24 \pm 50\%$	[164]

Note. The value of σ is given in units of 10^{-16} cm². The temperature dependence of the cross section is $\sigma_n^- \propto T^{-4}$. The results for which the temperature is not quoted were obtained at room temperature.

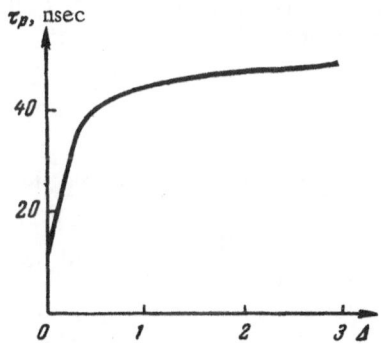

Fig. 7.4. Dependence of the hole lifetime in n-type silicon on the injection level Δ for $N_{Au} = 10^{15}$ cm⁻³.

strated that the lifetime calculated from the duration of the transient switching process is inversely proportional to the concentration of gold atoms. These measurements have been carried out on
samples with values of τ_p in the range 0.7–50 nsec.

The most reliable values of the capture cross section (Table
7.4) are those obtained recently by Fairfield and Gokhale [164].
These workers have determined the recombination parameters using two methods: from the photoconductivity decay during the flow
of a constant current through a silicon sample illuminated with
light modulated by means of a Kerr cell; from the transient switching characteristic of alloyed diodes prepared from such silicon.

Fairfield and Gokhale have introduced gold into silicon both
by diffusion and by doping the melt; the concentration of gold has
been determined by measuring the resistivity and by radioactivation analysis. The majority-carrier density in the investigated
samples has been found to lie within the limits 10^{15}–10^{17} cm^{-3} and
the concentration of gold atoms within 10^{14}–10^{16} cm^{-3}.

The values of the lifetime, obtained by the two methods mentioned above, have been found to differ by not more than 30%.

Since the acceptor and donor levels of gold lie close to the
middle of the forbidden band, the lifetime at low injection levels,
τ_0, is independent of the resistivity of silicon at all possible values
of the resistivity and is equal to $(v_p \sigma_p^- N_{Au})^{-1}$ for n-type silicon and
$(v_n \sigma_n^+ N_{Au})^{-1}$ for p-type silicon. Bearing in mind that the maximum
solubility of gold in solid silicon is 10^{17} cm^{-3} [161], we obtain the
following minimum lifetimes:

$$
\left.
\begin{aligned}
\tau_0 &= 90 \text{ nsec} \quad (n\text{-Si}), \\
\tau_0 &= 160 \text{ nsec} \quad (p\text{-Si})
\end{aligned}
\right\} \quad \text{according to } [164],
$$

$$
\left.
\begin{aligned}
\tau_0 &= 1000 \text{ nsec} \quad (n\text{-Si}), \\
\tau_0 &= 280 \text{ nsec} \quad (p\text{-Si})
\end{aligned}
\right\} \quad \text{according to } [162].
$$

The lifetime of nonequilibrium carriers at very high injection levels is governed by the recombination properties of both levels of gold and its value is given by the following expression,

which is obtained using Eq. (7.19) and the capture cross sections reported in [164]:

$$\tau_\infty = \frac{4.8 \cdot 10^7}{N_{Au}} \quad [\text{sec}].$$

(7.34)

The **dependence of the lifetime on the injection level**, calculated using Eq. (7.13), is shown in Fig. 7.4.

Literature Cited

1. M. F. Gardner and J. L Barnes, Transients in linear Systems Studied by the Laplace Transformation, Vol. 1, J. Wiley, New York (1942).
2. W. Shockley, Bell System Tech. J., 28(4):435-489 (1949).
3. S. G. Kalashnikov and N. A. Penin, Zh. Tekhn. Fiz., 25(6):1111-1123 (1955).
4. B. R. Gossick, Proc. Natl. Electr. Conf., 11:602-613 (1955).
5. É. I. Adirovich, Fiz. Tverd. Tela, 1(7):1115-1124 (1959).
6. Ya. A. Fedotov, Fundamentals of Physics of Semiconductor Devices, "Sov. Radio" (1963).
7. G. E. Pikus, Fundamentals of Theory of Semiconductor Divices, "Nauka" (1965).
8. R. H. Kingston, Proc. IRE, 42(5):829-834 (1954).
9. B. Lax and S. F. Neustadter, J. Appl. Phys., 25(9):1148-1154 (1954).
10. B. R. Gossick, Phys. Rev., 91(4):1012 (1953).
11. E. M. Pell, Phys. Rev., 90(2):278-279 (1953).
12. E. L. Steele, J. Appl. Phys., 25(7):916-918 (1954).
13. W. Shockley and W. T. Read, in: Electronic Semiconductor Devices [collection of Russian translations], IL (1953).
14. R. N. Hall, Phys. Rev., 87(2):387 (1952).
15. E. E. Rittner, Phys. Rev., 94(5):1161-1171 (1954).
16. V. I. Stafeev, Zh. Tekhn. Fiz., 28(8):1631-1641 (1958).
17. L. I. Baranov, Dissertation for Candidate's Degree, Saratov State University (1963).
18. A. A. Grinberg, Fiz. Tverd. Tela, 4(1):99-103 (1962).
19. A. N. Tikhonov and A. A. Samarskii, Equations of Mathematical Physics, Gostekhizdat (1951).
20. H. S. Carslaw, Introduction to the Mathematical Theory of the Conduction of Heat in Solids, Macmillan, London (1921); H. S. Carslaw and J. C. Jaeger, Conduction of Heat in Solids, 2nd ed., Clarendon Press, Oxford (1959).
21. V. A. Ditkin and A. P. Prudnikov, Operational Calculus in Two Variables and Its Applications, Fizmatgiz (1958).
22. V. A. Ditkin and P. I. Kuznetsov, Handbook on Operational Calculus, Gostekhizdat (1951).
23. V. A. Ditkin and A. P. Prudnikov, Handbook on Operational Calculus, "Vysshaya Shkola" (1965).
24. V. A. Ditkin and A. P. Prudnikov, Integral Transforms and Operational Calculus, Fizmatgiz (1961).

25. A. V. Lykov, A Theory of Heat Conduction, Gostekhizdat (1952).

26. Yu. R. Nosov, Radiotekhn. i Élektron., 6(2):313-320 (1961).

27. W. H. Ko, IRE Trans., ED-8(2):123-131 (1961).

28. I. P. Stepanenko, Izv. Vysshikh Uchebn. Zavedenii, Radiotekhnika, 4(2):175-184 (1961).

29. R. G. Shulman and M. E. McMahon, J. Appl. Phys., 24(10):1267-1272 (1953).

30. J. C. Henderson and J. R. Tillman, Proc. IEE, 104B(15):318-336 (1957).

31. R. H. Rediker and D. E. Sawyer, Proc. IRE, 45(7):944-953 (1957).

32. A. A. Grinberg and G. M. Avak'yants, Dokl. Akad. Nauk UzSSR, No. 7, pp. 31-36 (1956).

33. F. Finnegan, Vopr. Radiolokatsionnoi Tekhniki, No. 6(30), 95-103 (1955).

34. L. I. Baranov and M. S. Bekbulatov, Radiotekhn. i Élektron., 4(4):703-709 (1959).

35. T. R. Scott, Proc. IEE, 104B(15):333 (1957).

36. R. A. Smith, Semiconductors, University Press, Cambridge (1959).

37. Yu. R. Nosov, Silicon Pulse Diodes (Dissertation for Candidate's Degree), Moscow (1964).

38. A. I. Il'enkov, Radiotekhn. i Élektron., 8(5):830-833 (1963).

39. Yu. R. Nosov, Radiotekhn. i Élektron., 9(12):2122-2128 (1964).

40. Yu. R. Nosov, in: Semiconductor Devices and Their Applications (ed. by Ya. A Fedotov), No. 4, "Sov. Radio" (1960), pp. 3-38.

41. F. A. Zhuravel' and A. I. Il'enkov, Radiotekhn. i Élektron., 11(3):564-565 (1966).

42. M. I. Iglitsyn, Yu. A. Kontsevoi, and A. I. Sidorov, Zh. Tekhn. Fiz., 27(11): 2459-2460 (1957).

43. M. I. Iglitsyn, Yu. A. Kontsevoi, and A. I. Sidorov, Zh. Tekhn. Fiz., 27(11): 2461-2468 (1957).

44. M. I. Iglitsyn, Yu. A. Kontsevoi, and K. V. Temko, Radiotekhn. i Élektron., 5(3):508-513 (1960).

45. Yu. A. Kontsevoi, Dissertation for Candidate's Degree, Moscow (1961).

46. W. Webster, Proc. IRE, 42(6):914-920 (1954).

47. N. H. Fletcher, Proc. IRE, 45(6):862-872 (1957).

48. Yu. K. Barsukov, Zh. Tekhn. Fiz., 27(10):2252-2261 (1957).

49. Yu. K. Barsukov, Dissertation for Candidate's Degree, Leningrad Electrotechnical Institute (1959).

50. Yu. K. Barsukov, Fiz. Tverd. Tela, 1(4):602 (1959).

51. Yu. K. Barsukov, Fiz. Tverd. Tela, 1(11):1659-1667 (1959).

52. B. R. Gossick, J. Appl. Phys., 26(11):1356-1365 (1955).

53. B. R. Gossick, J. Appl. Phys., 27(8):905-910 (1956).

54. S. R. Lederhandler and L. J. Giacoletto, Proc. IRE, 43(4):477-483 (1955).

55. V. I. Gaman, Izv. Vysshikh Uchebn. Zavedenii, Fiz., No.(6):27-34 (1965).

56. T. E. Firle, M. E. McMahon, and J. E. Roach, Proc. IRE, 43(5):603-607 (1955).

57. T. E. Firle, IRE Wescon Conv. Rec., 1(3):90-95 (1957).

58. D. E. Rosenheim and A. G. Anderson, Proc. IRE, 45(2):212-219 (1957).

59. M. J. Callé, B. Dale, and C.A.P Fozel, Proc. IRE, 106B, Suppl. No.17:1138-1145 (1959).

60. I. A. D. Lewis and F. H. Wells, Millimicrosecond Pulse Techniques, Pergamon Press, London (1954); 2nd ed., Pergamon Press, London (1959).

61. V. G. Marants, In: Semiconductor Devices and Their Applications (ed. by Ya. A. Fedotov), No. 8, "Sov. Radio" (1962), pp. 137-174.
62. Yu. R. Nosov, Semiconductor Pulse Diodes, "Sov. Radio" (1965).
63. V. M. Shtein, Élektrosvyaz', No. 9, p.p. 12-19 (1959)
64. B. G. Vlaskin and V. M. Shtein, Élektrosvyaz', No. 9, pp. 68-71 (1960).
65. M. E. Glushkovskii, Izv. Vysshikh Uchebn. Zavedenii, Radiotekhnika, No. 1, pp. 3-17 (1963).
66. A. I. Il'enkov, Dissertation for Candidate's Degree, Novosibirsk (1963).
67. D. Yu. Éidukas, Dissertation for Candidate's Degree Kaunas (1962).
68. A. I. Il'enkov, Radiotekhn. i Élektron., 8(6):1019-1023 (1963).
69. A. I. Il'enkov, Izv. Sibirsk. Otdel. Akad. Nauk SSSR, No. 11, pp. 130-134 (1962).
70. A. I. Il'enkov, Izv. Sibirsk. Otdel. Akad. Nauk SSSR, Ser. Tekhn., No. 6, Part 2, pp. 27-31 (1963).
71. Yu. K. Barsukov. Zh. Tekhn. Fiz., 27(10):2262-2267 (1957).
72. Yu. K. Barsukov, Fiz. Tverd. Tela, 1(6):886-894 (1959).
73. D. Navon, R. Bray, and H. Y. Fan, Proc. IRE, 40(11):1342-1347 (1952).
74. Yu. R. Nosov and N. V. Postnikova, Radiotekhn. i Élektron., 9(12):2129-2132 (1964).
75. D. T. Stevenson and R. J. Keyes, Phys. Rev., 94(5):1416 (1954).
76. O. Curtis and B. R. Gossick, IRE Trans., ED-3(4):163-167 (1956).
77. J. B Arthur, A. F. Gibson, and J. B. Gunn, Proc. Phys. Soc. (London), 69B(7):697-704 (1956).
78. J. B. Arthur, A. F. Gibson, and J. B. Gunn, Proc. Phys. Soc. (London), 69B(7):705-711 (1956).
79. N. A. Penin, Radiotekhn. i Élektron., 2(8):1053-1061 (1957).
80. L. I. Baranov, Izv. Vysshikh Uchebn. Zavedenii, Fiz., No. 2, pp. 5-19 (1965).
81. Yu. F. Sokolov, Radiotekhn. i Élektron., 8(3):471-478 (1963).
82. Yu. A. Tkhorik, Radiotekhn. i Élektron., 10(3):574-576 (1965).
83. J. Halpern and R. H. Rediker, Proc. IRE, 46(6):1068-1076 (1958).
84. L. I. Baranov, Radiotekhn. i Élektron., 5(6):1002-1005 (1960).
85. I. M. Muratov and K. M. Kul'kin, Vysshikh Uchebn. Zavedenii, Fiz., No. 3, pp. 179-181 (1963).
86. I. M. Muratov, Radiotekhn. i Élektron., 10(12):2221-2225 (1965).
87. M. Byczkowski and J. R. Madigan, J. Appl. Phys., 28(8):878-881 (1957).
88. I. M. Muratov, Radiotekhn. i Élektron., 10(1):167-170 (1965).
89. V. I. Gaman and V. M. Kalygina. Izv Vysshikh Uchebn. Zavedenii, Fiz., No. 5, pp. 77-79 (1965).
90. N. A. Penin and K. V. Cherkas, Radiotekhn. i Élektron., 3(12):1495-1500 (1958).
91. J. R. Tillman and H. Yemm, Phil. Mag., 41(323):1281-1283 (1950).
92. L. A. Meacham and S. E. Michaels, Phys. Rev., 78(2):175-176 (1950).
93. J. H. Wright. Proc. IRE, 40(2):232 (1952) [abstract only].
94. D. J. Crawford and H. F. Heath, Proc. IRE, 40(2):232 (1952).
95. M. C. Waltz, Proc. IRE, 40(11):1483-1487 (1952).
96. A. C. Sim, J. Electron. Control, 3(2):139-159 (1957).
97. G. D. Glebov (ed.), Manufacture of Semiconductor Devices, Oborongiz (1962).

98.   S. N. Ivanov, Radiotekhn. i Élektron., 8(6):1074-1076 (1963).

99.   S. N. Ivanov, N. A. Penin, N. E. Skvortsova, and Yu. F. Sokolov, Physical Basis of Operation of Semiconductor Microwave Diodes, "Sov. Radio" (1965).

100. H. L. Armstrong, J. Appl. Phys., 27(4):420-421 (1956).

101. H. L. Armstrong, Proc. IRE, 45(5):696-697 (1957).

102. H. L. Armstrong, E. D. Metz, and I. Weiman, IRE Trans.,ED-3(2):86-93 (1956).

103. S. A. Eremin, O. K. Mokeev, and Yu. R. Nosov, Semiconductor Charge-Storage Diodes and Their Applications, "Sov. Radio" (1966).

104. C. T. Sah, R. H. Noyce, and W. Shockley, Proc. IRE, 45(9):1228-1243 (1957).

105. J. L. Moll, S. Krakauer, and R. Shen, Proc. IRE, 50(1):43-53 (1962).

106. V. I. Gaman, Izv. Vysshikh Uchebn. Zavedenii, Fiz., No. 2, p.p. 73-77 (9165).

107. D. P. Kennedy, IRE Trans., ED-9(2):174-182 (1962).

108. S. Y. Muto and S. Wang. IRE Trans., ED-9(2):183 (1962).

109. Y. Kanai, J. Phys. Soc. Japan, 10(8):719-720 (1955).

110. E. Spenke, Z. Angew. Phys., 10(2):65-88 (1958).

111. I. Ladany, IRE Trans., ED-7(4):303-310 (1960).

112. E. Rocher, Z. Angew. Phys., 14(6):347-352 (1962).

113. S. P. Sinitsa, Radiotekhn. i Élektron., 7(8):14-27 (1962).

114. M. I. Iglitsyn and V. I. Fistul', Dokl. Akad. Nauk SSSR, 149(3):577-579 (1963).

115. G. B. Abdullaev, Z. A. Iskanderzade, and É. A. Dzhafarova, Radiotekhn. i Élektron., 10(4):776-778 (1965).

116. T. Einsele, Z. Angew. Phys., 4(5):183-185 (1952).

117. R. Bray and B. R. Gossick, Phys. Rev., 91(4):1011-1012 (1953).

118. G. Kohn and W. Nonnenmacher, Arch. Elektr. Übertrag., 8(11):561-564 (1954).

119. W. Guggenbühl, Arch. Elektr. Übertrag., 10(11):483-485 (1956).

120. G. Kohn, Arch. Elektr. Übertrag., 9(5):241-245 (1955).

121. T. Misawa, J. Phys. Soc. Japan, 12(8):882-890 (1957).

122. C. G. Dorn, IRE Trans., ED-3(3):153-156 (1956).

123. W. Heinlein, Arch. Elektr. Übertrag., 11(10):387-396 (1957).

124. K. Kano and H. J. Reich, IEEE Trans., ED-11(11):515-523 (1964).

125. Y. F. Chang, J. Appl. Phys., 34(7):2056-2060 (1963).

126. H. L. Armstrong, IRE Trans., ED-4(2):111-113 (1957).

127. G. M. Guro, Zh. Tekhn. Fiz., 33(1):158-165 (1957).

128. É. I. Adirovich and G. M. Guro, Dokl. Akad. Nauk SSSR. 108(3):417-420 (1956).

129. D. J. Sandiford, Phys. Rev., 105(2):524 (1957).

130. S. M. Ryvkin, Photoelectric Effects in Semiconductors, Consultants Bureau, New York (1964).

131. Yu. A. Kontsevoi, Fiz. Tverd. Tela, 1(8):1289-1293 (1959).

132. C. T. Sah and W. Shockley, Phys. Rev., 109(4):1103-1115 (1958).

133. N. G. Zhdanova, S. G. Kalashnikov, and A. I. Morozov, Fiz. Tverd. Tela, 1(4):535-544 (1959).

134. W. van Roosbroeck and W. Shockley, Phys. Rev., 94(6):1558-1560 (1954).

135. J. A. Hornbeck and J. R. Haynes, in: Problems in Physics of Semiconductors (ed. by V. L. Bonch-Bruevich), IL (1957), pp. 167-203.

136. H. A. Gebbie, M. Nisenoff, and H. Fan, Phys. Rev., 91(1):230 (1954) [abstract only].

137. F. M. Berkovskii, S. M. Ryvkin, and N. B. Strokan, Fiz. Tverd. Tela, 3(1):230-235 (1961).

138. F. M. Berkovskii, S. M. Ryvkin, and N. B. Strokan, Fiz. Tverd. Tela, 3(11):3535-3537 (1961).

139. É. I. Adirovich, A. N. Gubkin, and B. D. Kopylovskii, Tr. Fiz. Unet. Akad. Nauk SSSR, 20:126-171 (1963).

140. V. E. Lashkarev, Izv.Akad. Nauk SSSR, Ser. Fiz., 16(2):186-201 (1952).

141. G. M. Guro, Usp. Fiz. Nauk, 72(4):711-740 (1960).

142. S. G. Kalashnikov, Proceedings of the Fifth International Conference on Physics of Semiconductors, Prague, 1960, publ. by Academic Press, New York (1961). pp. 241-252.

143. J. Johnson and H. Levinstein, Phys. Rev., 117(5):1191-1203(1960).

144. K. D. Glinchuk, E. G. Miselyuk, and N. N. Fortunatova, Fiz. Tverd. Tela, 1(9): 1345-1350 (1959).

145. T. P. Vogel, J. R. Hansen, and M. Garbuny, J. Opt. Soc. Am., 51(1):70-75 (1961).

146. L. J. Neiringer and W. Bernard, Phys. Rev. Letters, 6(9):455-457 (1961).

147. J. L. Rupprecht, Procceedings of the Fifth International Conference on Physics of Semiconductors, Prague, 1960, publ. by Academic Press, New York (1961). pp. 282-286.

148. I. V. Karpova, Dissertation for Candidate's Degree, Inst. Radioeng. i Élektron. Akad. Nauk SSSR (1963).

149. L. Ya. Pervova, Radiotekhn. i Élektron., 6(10): 1745-1748 (1961).

150. V. G. Alekseeva, I. V. Karpova, and S. G. Kalashnikov, Fiz. Tverd. Tela. 3(3): 964-971 (1961).

151. I. L. Kurova, S. G. Kalashnikov, and N. D. Tyapkina, Fiz. Tverd. Tela. 4(6): 1503-1509 (1962).

152. P. G. Eliseev and S. G. Kalashnikov, Fiz. Tverd. Tela, 5(1):320-326 (1963).

153. N. G. Zhdanova and V. G. Alekseeva, Fiz. Tverd. Tela, 5(2):546-551 (1964).

154. I. A. Kurova and V. V. Ostroborodova, Fiz. Tverd. Tela, 7(3):683-686 (1965).

155. V. V. Ostroborodova, Fiz. Tverd. Tela, 7(2):610-618 (1965).

156. A. S. Syed, Can. J. Phys., 40(2):286-288 (1962).

157. W. W. Tyler, J. Phys. Chem. Solids, 8(1):59-65 (1959).

158. H. Kodera, J. Japan. Appl. Phys., 3(7):369-376 (1964).

159. Yu. R. Nosov and L. V. Gubyrin, Radiotekhn. i Élektron., 10(3):570-572 (1965).

160. A. P. Klimenko and Yu. A. Tkhorik, Ukr. Fiz. Zh., 10(2):238-239 (1965).

161. C. B. Collins, R. O. Carlson, and C. J. Gallagher, Phys. Rev., 105(4):1168-1173 (1957).

162. G. Bemski, Phys. Rev., 111(6):1515-1518 (1958).

163. W. D. Davis, Phys. Rev., 114(4):1006-1008 (1959).

164. J. M. Fairfield and B. V. Gokhale, Solid State Electronics, 8(8):685-691 (1965).

165. A. E. Bakanowski and J. H. Forster, Bell System Tech. J., 39(1):87-104 (1960).

166. Yu. A. Tkhorik, Transient Processes in Semiconductor Pulse Diodes, Izd. "Tekhnika," Kiev (1966).

167. W. Shockley, Phys. Rev., 125(5):1570-1576 (1962).

Index

B

barrier capacitance, effect on transient
    processes 45
base (definition) 4
built-in field
  behavior during recovery phase 171-175
  behavior during reverse phase 175-178
  definition 162
  forward-biased diodes 165-171
  general theory 160-165

C

carrier recombination
  effect on transient processes 197-208
  germanium, Au-doped, effect on tran-
    sient processes 215-222
  silicon, Au-doped, effect on transient
    processes 223-225
carrier trapping, effect on transient pro-
    cesses 209-213
contacts
  nonrectifying, classification 97-99
  ohmic, noninjecting 101-102
  ohmic, recombination-type 99-101

D

Debye screening length 10
delayed switching
  experimental results (review) 88-92
  general theory 35-38
dielectric relaxation time 10
diffusion equation, solution for low in-
    jection levels 10-21

E

emitter (definition) 4

experimental results (review)
  planar semi-infinite diodes
    delayed switching 88-92
    postinjection emf decay 92-95
    recovery phase 85-88
    reverse phase 82-85
  point-contact diodes 153-159

G

germanium, effect of recombination on
    transient processes 215-222

H

hemispherical-junction diodes, transient
    processes 191-196
hole distribution
  planar diode with thin base 96-107
  under forward bias 21-24

I

impressed hole density (definition) 12
injection efficiency of contact 11

L

Laplace—Carson transformation 16

M

Maxwellian relaxation time see dielec-
    tric relaxation time
microalloyed point-contact diodes 155-
    159

N

nonrectifying contacts, classification 97-99

O

observation of transients in diodes
  experimental errors 76-82

observation of transients in diodes
(continued)
  measuring apparatus 69-76
ohmic contacts
  noninjecting 101-102
  recombination-type 99-101
P

planar diodes
  finite base, transient processes 183-191
  thin base 96-129
    with finite limiting resistance 118-127
    without limiting resistance 107-118
  semi-infinite base 25-95
    with finite limiting resistance 38-53
    with infinite limiting resistance 53-66
    without limiting resistance 25-38
point-contact diodes 130-159
  comparison with planar diodes 151-153
  experimental results (review) 153-159
  ideal model 131-141
  microalloyed diodes 155-159
  transient processes 141-153
postinjection emf
  definition 54
  experimental results (review) 92-95
R

radio pulses 1
recombination see carrier recombination
recovered charge (definition) 27
recovery phase
  definition 39
  experimental results (review) 85-88
relaxation time, dielectric 10
reverse phase
  definition 39
  experimental results (review) 82-85
S

silicon, effect of recombination on tran-
    sient processes 223-225
small-signal approximation
  definition 3
  transient characteristics of diodes 66-69
steady-state (definition) 1

step function 2
stored charge
  definition 27
  thin-base diodes 104-107
switching processes see also transient pro-
    cesses
  delayed switching
    experimental results (review) 88-92
    general theory 35-38
  general equations 4-10
  planar diodes
    finite base 183-191
    semi-infinite base 25-95
    thin base 96-129
  point-contact diodes 130-159
  postinjection emf 54, 92-95
T

transient processes
  during passage of forward current 179-
    196
  germanium, Au-doped, effect of re-
    combination 215-222
  hemispherical-junction diodes 191-196
  influence of carrier recombination 197-
    208
  influence of carrier trapping 209-213
  observation methods
    experimental errors 76-82
    measuring apparatus 69-76
  planar diodes with finite base 183-191
  planar diodes with semi-infinite base
    25-95
    with finite limiting resistance 38-53
    with infinite limiting resistance 53- ·
    66
    without limiting resistance 25-38
  planar diodes with thin base 96-129
    with finite limiting resistance 118-
    127
    without finite limiting resistance 107-
    118
  point-contact diodes 141-153
  silicon, Au-doped, effect of recombina-
    tion 223-225

transient state (definition) 1-2

transfer function 2

trapping see carrier trapping

V

video pulses 1